Introduction to
Molecular Medicine

Third Edition

Dennis W. Ross, M.D., Ph.D.
Department of Pathology
Forsyth Medical Center and
Wake Forest University
School of Medicine
Winston-Salem, North Carolina

Introduction to
Molecular Medicine
Third Edition

Illustrations by David Pounds

With 58 Illustrations

 Springer

Dennis W. Ross, M.D., Ph.D.
Department of Pathology
Forsyth Medical Center
3333 Silas Creek Parkway
Winston-Salem, NC 27103, USA
and
Clinical Professor, Pathology
Wake Forest University
School of Medicine
Winston-Salem, NC 27103, USA
dwross@novanthealth.org

Library of Congress Cataloging-in-Publication Data
Ross, D.W. (Dennis W.)
    Introduction to molecular medicine / Dennis W. Ross.—3rd ed.
      p. ; cm.
    Includes bibliographical references and index.
    ISBN 0-387-95372-8 (s/c : alk. paper)
    1. Medical genetics.   2. Molecular biology.   3. Pathology, Molecular.   I. Title.
[DNLM: 1. Genetics, Medical.   2. Hereditary Diseases—genetics.   3. Molecular Biol-
ogy.   4. Neoplasms—genetics.   QZ 50 R823i 2002]
   RB155 .R78 2002
   616′ .042—dc21   2001054921

ISBN 0-387-95372-8   Printed on acid-free paper.

Printed in the United States of America.   (IMP/MVY)

9 8 7 6 5 4

springeronline.com

Preface to the Third Edition

This book continues the story of a new field called molecular medicine. The first edition appeared 10 years ago in a previous millennium. At that time cloning a gene was cutting edge research; now gene cloning is a high school science fair project. The complete sequencing of the human genome was finished last year; cloning a human is a likely (and very controversial) event for 2002. The tools of molecular medicine have become very powerful and automated; DNA on a chip and microarrays allow us to probe large parts of the genome and its function. The speed of sequencing DNA has increased about 100,000-fold over the decade since the first edition of this book was published. Following the great improvements in technology, DNA and molecular research have a rapidly deepening impact on human medicine. Molecular therapies including DNA vaccines, antisense, and gene transplantation are undergoing clinical trials.

This third edition is almost entirely new. All of the figures and about 90% of the text are new. The message is the same. I present the discoveries, basic scientific concepts, and sense of excitement that surrounds what is clearly a revolution in medicine. Do not be put off by technical detail and jargon; it will all be explained. Concepts in boldface type are defined in the glossary. Each chapter begins with an overview and ends with a summary. This book is not a comprehensive monograph; I stick to the basic concepts as demonstrated by important clinical examples. The goal is to prepare us for the very great changes that will result from our decoding and understanding of the human genome and the molecular basis of life.

Dennis W. Ross, M.D., Ph.D.
Winston-Salem, North Carolina

Preface to the Second Edition

In the four years since the first edition of this book was published, the molecular revolution has continued. DNA has been named by *Time* magazine as the Molecule of the Year, a Nobel Prize has been awarded to a young man for the invention of the polymerase chain reaction, and television viewers have learned of the DNA fingerprint. Molecular technology in medicine is increasing. The availability of DNA probes for cancer susceptibility is stressing our system of insurance, testing our ideas about medical ethics, and teaching us new things about cancer. In this edition, I have added a number of new sections, as well as a new chapter. New examples of molecular medicine have been added to demonstrate current applications of this technology. The basic concepts of molecular biology remain the basis for the first three chapters of the book. The excitement surrounding molecular medicine that I mentioned in the preface to the first edition continues. It is now tinged with a touch of awe and a little bit of fear at the changes that recombinant DNA technology have brought to our society.

Dennis W. Ross, M.D., Ph.D.
Winston-Salem, North Carolina

Preface to the First Edition

This book describes the discoveries that have created a field called molecular medicine. The use of recombinant DNA technology in medical research and most recently in medical practice constitutes a revolutionary tool in our study of disease. Probing the human genome is rapidly becoming as routine as looking at cells under the microscope. The cloning of a new gene is now a common occurrence, newspapers report. Recombinant DNA technology, like the invention of the microscope, shows us a world of detail richer than we might have imagined.

This book presents the discoveries, basic scientifiic concepts, and sense of excitement that surround the revolution in molecular medicine. The scientific basis of molecular medicine is explained in a simple and direct way. The level of technical detail, however, is sufficient for the reader to appreciate the power of recombinant DNA technology. This book is clinically oriented throughout. All of the examples and applications are related to medical discoveries and new methods of diagnosis and therapy. A few subjects within molecular medicine are examined in more detail to allow the reader to become aware of the strengths and shortcomings of a molecular approach to disease. I do not hide the incomplete understanding that still surrounds many of the recent discoveries in molecular medicine.

I intend to demonstrate the concepts of molecular medicine in this book by showing examples from all branches of medicine. I include, for instance, infectious diseases, genetic disorders, and cancer. However, I am not trying to be comprehensive in examining all areas of molecular medicine. So many discoveries are made each week in this field that it is not yet possible to draw them together in a comprehensive volume. My goal is to help the reader understand what the future may hold as well as the most important current applications.

Dennis W. Ross, M.D., Ph.D.
Winston-Salem, North Carolina

Acknowledgments

I wish to thank many people who have made useful suggestions and sustained warm discussions that have improved this third edition. Medical students have taught me, while I was in the apparent act of trying to teach them. Special thanks to Angela Bartley and Tehnaz Parakh. The cancer patients that I have worked with have demonstrated to me the need to turn research into medicine. My medical colleagues including: Malcolm Brown, David Collins, Joseph Dudley, Tommy Simpson, and Edward Spudis have tolerating my practicing exposition on them. Dr. Gordon Flake deserves special thanks for long continuing assistance to this book and its sister *Introduction to Oncogenes and Molecular Cancer Medicine* (Springer 1998). Laura Gillan, my editor at Springer, made me do it. At a time when I was not sure I could encompass all that was happening, she said, "just keep it simple" and I tried. All of the illustrations in this third edition are new. David Pounds, a medical illustrator, has taken all of the figures done by me in the earlier editions and re-interpreted or replaced them with drawings that are both more artistic and more correct. Finally, I need to thank my wife of 32 years, Suzie Ross, for the enduring encouragement and very real support of my writing and my career.

Dennis W. Ross, M.D., Ph.D.
Winston-Salem, North Carolina

Contents

I.   BASICS OF MOLECULAR BIOLOGY

Overview • The Genetic Message • Information Content of the Genome • Open Reading Frame, ORF • Genome versus Library • Anatomy of a Gene • BRCA1 • C-MYC • INS • Junk DNA • Gene Maps • Physical Organization of the Genome • Basic Genetics • Alleles and Inheritance • Polygenic Traits • Human Genome Project • Aside: What the Genome Does Not Code For • Summary

Overview • DNA Replication • Mitosis and Meoisis • DNA Repair • Mutations • Point Mutation • Frame Shift Mutation • Chromosomal Translocation • Aside: Teleomeres and Aging • DNA to Protein • Proteomics • Gene Expression • Insulin • Cell Signaling Pathway • Summary

Overview • Restriction Enzymes • Southern, Northern and Western Blots • DNA Hybridization • Polymerase Chain Reaction • DNA on a Chip • Expression Arrays • Antisense Oligonucleotides • Knockout Mice • Gene Vectors • Cloning Vectors and Gene Therapy of Cystic Fibrosis • Cloning • Cloning a Gene • Cloning a Human • Choice of Donor Cell • In Vitro Fertilization • Re-programming • Genetic Engineering of the Clone • Fertilization • Implantation and Birth • Aside: Human Spare Parts • Summary

II. MOLECULAR APPROACH TO DISEASE

examples • Hemophilia and Factor IX Replacement • Immunotherapy of Lymphoma/Leukemia • Suicide Gene Therapy of Tumors • Graft versus Host Therapy of Leukemia • Molecular Production of Drugs • Antisense • Recombinant Drugs • Pharmed Drugs • Mini-examples • CEA Antigen in Cereals • Aside: Franken Foods • Biomaterials and Tissues • Embryonic Stem Cells

PART I

Basics of Molecular Biology

1   The Human Genome

Overview

A gene is a piece of DNA that encodes information that is part of a cell's permanent structure and is copied into daughter cells at division. The human genome contains 3 billion base pairs that code for about 35,000 genes. The anatomy of a gene shows that it contains many regulatory elements that control gene expression as well as specifying the amino acids that make up its protein. We will look in detail at a few sample genes, like the pro-insulin INS gene, in order to prepare for clinical problems such as the treatment of diabetes by genetic engineering. Basic genetics has itself been greatly amplified by our molecular knowledge of the human genome. Simply put; things are more complex than Mendel's wrinkled peas. Polygenic traits that effect major diseases such as atherosclerosis depend on many genes and the interaction of these genes with outside influences. Finally, even as we realize that we have conquered the human genome, we must recognize that there is much that is important for which the genome does not code.

The Genetic Message

Within every cell in the body is the information necessary to specify the physical structure of an entire human being. The information is the genome, written in DNA as a string of 3 billion letters using an alphabet that consists of only four bases: A—adenine, T—thymine, C—cytosine, and G—guanine. The DNA content of a single cell is 7.1 picograms (10^{-12} grams); present within the cell nucleus as a linear strand 2 meters longs by 0.2 nanometers wide (Table 1.1).

I have found it useful to consider several analogies to grasp the size and complexity of the human genome. To begin, imagine the DNA molecule scaled up to the size of a railroad track with the rails as the backbone of the DNA helix and the cross-ties as the base pairs: A-T or G-C.

Table 1.1. Parameters of the human genome.

Number of base pairs (bp) = 6 x 10^9 (diploid genome)
  1% to 5 % of these base pairs code for protein, the rest are regulatory elements
  and "junk"
Number of genes = 35,000 for haploid genome
Size of a gene = from 1,000 to 200,000 bp; typical gene 30,000 bp
24% of the genome is in introns
75% of the genome is between genes
50% of genome consists of repeat sequences
Differences in genome from one person to another: 0.1% or 1 per 1000 bp
Number of single nucleotide polymorphisms (SNPs) = 2–2.5 x 10^6
Differences in genome from human to chimpanzee: 2% or 20 per 1,000 bp
Number of chromosomes = 22 pairs plus XX or XY
Size of chromosomes: largest (no. 1) = 279 Mb; smallest (no. 22) 48 Mb
1 centimorgan (cM) = about 1 million base pairs (Mb)
Length of DNA as a single linear strand = 2 m
Width of DNA double-helix molecule = 2 × 10^{-9} m
Helix rotates 360 degrees every 10 bp = 3.4 × 10^{-9} m
Weight of DNA in one cell = 7.1 × 10^{-12} g

This railroad track would be a million miles long. The cross-ties carry the genetic information. Walking up one rail might show the sequence ATGGGTCTG; a friend walking past you down the other rail will see the complementary bases TACCCAGAC as shown in Fig. 1.1. Sequencing the human genome means traveling a million miles and writing down the letters seen on 3 billion cross-ties. Twenty years ago a few dozen researchers were scattered over those million miles covering a distance of perhaps 100 base pairs (bp) a day. Ten years ago as the Human Genome Project began, about a thousand research groups were able to move at speeds up to 10,000 bp a day. Now with modern sequencing techniques and DNA on a chip technology, we approach 1,000 bp a second in our decoding. That's an amazing speed, but still about 10 times slower than the speed at which a cell can replicate its own DNA. During the cell division cycle, the entire diploid genome, two copies of the DNA, one on each pair of chromosomes, is replicated within 12 hours.

Information Content of the Genome

A **gene** is a piece of DNA that encodes a packet of information that is part of a cell's permanent structure and is copied into daughter cells at division. The three essential elements of this definition are *DNA, information,* and *copied.* A gene encodes a specific linear sequence of amino acids assembled on the polyribosomes of the cells. Figure 1.2 demonstrates a genetic message *encoded* in DNA, *transcribed* into messenger RNA, and *translated* into a protein. The events that control the expression of the genetic message (the subject of Chapter 2) control the function of the cells in the body by modulating the synthesis of proteins.

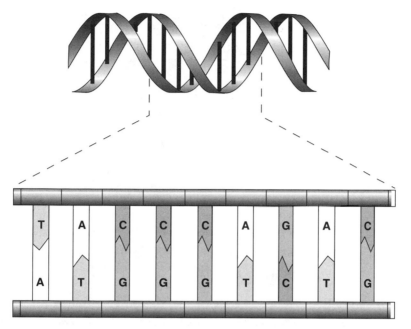

Figure 1.1. The DNA molecule is a double-stranded helix. Genomic information is coded on the base pairs, A-T and C-G, that cross-link the two strands.

The genetic message is grouped into three-letter words called **codons**. Each codon specifies one of the 20 possible amino acids that are the building blocks of all proteins. Table 1.2 lists these 20 amino acids (including their three-letter and one-letter abbreviations) and the corresponding codons. Note that the base uracil is used in place of thymine in RNA. Thus genetic sequences when written in RNA contain a U everywhere there was a T in the corresponding DNA. This genetic code is used for all life on the earth![1] The "stop" codons UAA, UAG, and UGA do not specify an amino acid but serve as periods to end a message. The 3 billion base pairs of the genome code for an estimated 35,000 genes. It is important to get an idea of how much information this is; again, an analogy is helpful. The 3 billion letters of the genome are about the same as the number of letters in all the books of a good medical school library. In the library, letters are used to specify words of varying length grouped into sentences with punctuation. These in turn are further organized into paragraphs, chapters, and books. In the genome, there are only three-letter codons with punctuation limited to start and stop signals.

---

[1]To be very exact, there are a few rare microorganisms and other strange living things that do not use the exact coding as reflected in Table 1.2. I warn the reader now that there are always exceptions and variations to every rule. Sometimes I will not point these out; sometimes I might even be in error and not aware of the exceptions.

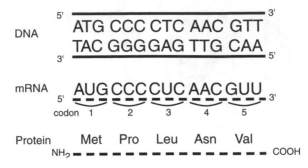

Figure 1.2. A genetic message begins as a double-stranded DNA molecule that serves as the template for messenger RNA. The mRNA, in groups of three bases to a codon, directs the order of amino acids in protein.

Opening Reading Frame, ORF

When we look at a long DNA sequence, the punctuation is not readily visible. We know that the bases are read as three-letter codons, but where does a codon start? If you think about it, there are three possible reading frames, only one of which is correct. Consider the following example, written in three-letter English words instead of three-base codons:

FTHEOWLANDCATATECODONEANDALLEG

An open reading frame that produces a sensible message begins at the second letter. Reading frames that begin at the first and third letter produce messages that make no sense.

Finding the correct reading frame in a DNA sequence is a matter of searching for a long stretch of codons without stop codons. Only within a gene, and only if the correct frame of reference is used, will a long stretch of DNA sequences be found in which none of the three stop codons occurs. Everywhere else, within incorrect reading frames and in parts of the DNA that are outside of gene sequences, the stop codons will be encountered by chance every 30 or so bases. A DNA sequence that does not contain stop codons is called an **open reading frame (ORF)**. When researchers uncover a new DNA sequence, they get quite excited if they find an ORF; that means they are within a new gene rather than in a noncoding portion of the genome.

Genome versus Library

In a library, books are placed on shelves. Genes are on 23 pairs of chromosomes. There is one copy of each gene on each chromosome of the pair. However, the copies are not necessarily identical. The genome

Table 1.2. The genetic code.

AA	Symbols	Amino acid	Codons
A	Ala	Alanine	GCA GCC GCG GCU
C	Cys	Cysteine	UGC UGU
D	Asp	Aspartic acid	GAC GAU
E	Glu	Glutamic acid	GAA GAG
F	Phe	Phenylalanine	UUC UUU
G	Gly	Glycine	GGA GGC GGG GGU
H	His	Histidine	CAC CAU
I	Ile	Isoleucine	AUA AUC AUU
K	Lys	Lysine	AAA AAG
L	Leu	Leucine	UUA UUG CUA CUC CUG CUU
M	Met	Methionine	AUG (Met is also a START codon)
N	Asn	Asparagine	AAC AAU
P	Pro	Proline	CCA CCC CCG CCU
Q	Gln	Glutamine	CAA CAG
R	Arg	Arginine	AGA AGG CGA CGC CGG CGU
S	Ser	Serine	AGC AGU UCA UCC UCG UCU
T	Thr	Threonine	ACA ACC ACG ACU
V	Val	Valine	GUA GUC GUG GUU
W	Trp	Tryptophan	UGG
Y	Tyr	Tyrosine	UAC UAU
-	—	STOP codons	UAA UGA UAG

supplies information by transcribing a copy of DNA into messenger RNA (mRNA). The mRNA serves as a blueprint for construction of a protein. A few minutes after use, the transcript is destroyed. A cell can find any gene it needs within seconds and synthesize a new protein within minutes.

At this point, the library analogy weakens profoundly (Table 1.3). In a library, books can be identified through a card catalog or computer database. For the genome, the search mechanism is mostly unknown. In a library, books are grouped on shelves by subject. Genes do not seem to be grouped at all on the chromosomes; their location is seemingly random. There are some gene clusters such as the β-globulins on chromosome 11, the major histocompatibility transplant antigens on 6, and the B-lymphocyte immunoglobulin heavy chains on 14. In general, however, we do not see any overall physical organization of genes by function.

The greatest difference between a library and the human genome is that the books in a library are written by humans while the genome is a product of natural evolution. The information in a library changes rapidly. We read several books from the library, have an idea, and write a new book that is added back to the library. The cell cannot add to its storehouse of information based on its experience. The genome evolves slowly by natural selection, recombination of genes from parents as part of sexual reproduction, and mutation.

Table 1.3. Information storage—library versus genome.

	Library	Human genome
Size	Thousands of books	35,000–50,000 genes
Encoding	26-letter alphabet Words of varying length Blank spaces and punctuation	Four-letter alphabet Three-letter codons Start and stop codons control sequences
Organization	On shelves by subject	Chromosomal location seemingly at random
Search mechanism	Card catalog or database	Unknown
Circulation of information	Books are removed	Genes are transcribed
Source	Written by humans	Natural evolution (so far)

However, at the start of the new millennium, the rules for the human genome are changing. Now we humans are quite capable of writing within our own DNA and adjusting the information within our own internal libraries. To me, the ability to read and write our genetic heritage is the start of a revolution in human culture, the outcome of which will not be clear for some time. We will discuss this again in Chapters 3 and 6 when we consider genetic engineering.

Anatomy of a Gene

Medical students in the past had both gross and microscopic anatomy as grueling subjects; current students must add molecular anatomy to their curriculum. With 35,000 human genes, this could be a very difficult subject, especially to memorize. However, several basic principles of gene anatomy are sufficient to teach what is important. Genes are organized into segments called **exons** that are separated by **introns**. The exons are transcribed in mRNA and then translated into proteins. The base pairs within the introns do not code for protein. A great deal of the human genome is made up of introns, and other noncoding DNA sequences; almost 90% of the 3 billion base pairs in the genome do not contribute to the amino acid sequence of any protein. Most of this DNA is "junk" with an unclear purpose, as we will discuss later. Let's begin our study of gene anatomy with a few examples.

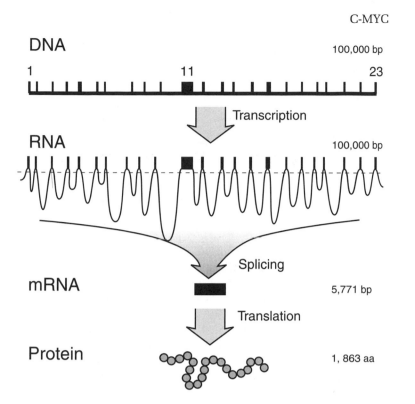

Figure 1.3. The *BRCA1* gene DNA sequence information is processed into protein by the intermediate steps of (1) transcription into RNA, (2) splicing out introns resulting in a much shorter mRNA, and (3) translation into protein.

BRCA1

When a gene is transcribed into RNA, the entire sequence is copied. However, as the mRNA molecule is processed, all the bases corresponding to intron sequences are spliced out. Figure 1.3 shows a diagram of the *BRCA1* breast cancer susceptibility gene. This gene is a large one with 23 exons spanning 100,000 bp. Figure 1.3 demonstrates the transcription of 100,000 bp DNA to RNA. The intron sequences are spliced out resulting in a compact mRNA molecule of 5,771 bases, that translate to a protein of 1,863 amino acids (aa).

C-MYC

Figure 1.4 is a more detailed diagram of another gene, the *C-MYC* growth control oncogene. This is a page from the anatomy books of the future. At the top of Fig. 1.4 is an icon of chromosome 8 as it would appear in a Giemsa-stained metaphase karyotype. The light- and dark-

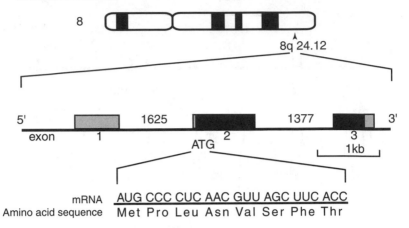

Figure 1.4. The anatomy of the *C-MYC* oncogene on chromosome 8 (arrowhead) shows a grouping into three exons. The initiation sequence ATG starts the mRNA template coding for the amino acid sequence of the protein.

staining bands on the short (p) and long (q) arms of the chromosome are at the limit of visibility under the microscope. An arrowhead indicates the submicroscopic location of *C-MYC* on the long arm at 8q24.12 The lower panel shows the genomic structure of *C-MYC*. This is a small gene, consisting of three exons and spanning only 5,000 bp. The spacing between exons 1 and 2 of the *C-MYC* gene is 1,625 bp and between exons 2 and 3, 1,377 bp. The first exon of *C-MYC* is not translated but contains a number of regulatory elements. Figure 1.5 is an annotated genetic code of *C-MYC*. The start of exon 2 begins as an open reading frame. The translation of the c-myc protein begins 16 bp later with the start codon ATG specifying methionine. The open reading frame of exon 3 ends with the stop codon TAA. At the end of the third exon are coding sequences for a polyA tail common to the end of mRNA molecules. The *C-MYC* mRNA is approximately 2,300 bp long and codes for a protein of 453 aa.

INS

Another simple gene, one of the very first genes sequenced, is the proinsulin gene, *INS*. Figure 1.6 shows that this gene has three exons that code for a 450-bp mRNA. The first exon will be transcribed but not translated into protein; it is an untranslated region (UTR). The second exon codes for a small signal peptide, the A chain, and part of the connecting piece C. The third exon codes for the rest of the C piece and the B chain. After the protein is assembled on the ribosomes in the cytoplasm, the connecting piece is cleaved out and the two chains, A and B, become covalently joined by two disulfide bonds. This is a

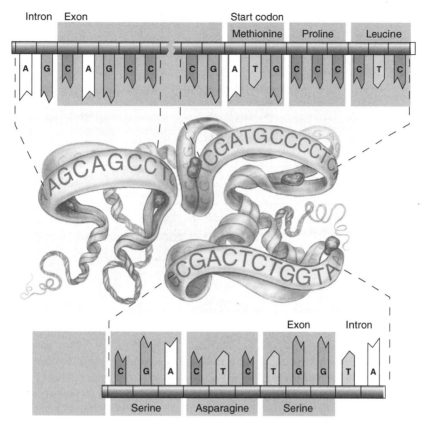

Figure 1.5. An annotated diagram of the *C-MYC* gene sequence shows functional features such as intron-exon boundaries, along with initiation and termination codons.

post-translational modification of a protein that occurs when the proinsulin molecule is created in the right environment. The final assembly of the molecule occurs in steps that are not encoded in the DNA. Most genes contain many more than three exons and have much more complicated splicing and posttranslational modification. Large genes with multiple exons can be read in a variety of ways, producing different proteins based on alternative transcriptions of the gene. *INS* is about as simple as it gets.

Another feature of importance, **alleles** of the *INS* gene, is also shown in Fig. 1.6. Two common variations of the genetic sequence that are seen in different individuals are denoted the alpha and beta alleles. The figure shows the bases that vary for each of these two alleles. We will discuss alleles later in this chapter. Basically an allele is an alternate copy of a gene that shares a common locus (place on the genome).

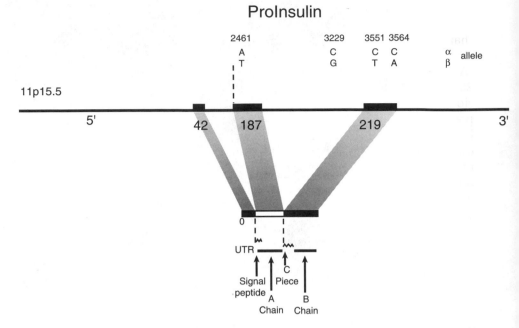

Figure 1.6. The proinsulin gene (*INS*, OMIM entry no. 176730) consists of three short exons that code for a precursor protein that is modified within the cell to form biologically active insulin.

Figure 1.6 is a map of the *INS* gene. It tells us something but not all we want to know. In Chapter 2 we will discuss regulation of insulin expression. Figures 2.4 and 2.5 will show how the *INS* gene is regulated to maintain a homeostatic control of glucose metabolism. In Chapter 5 we will consider how to genetically engineer this gene (see Fig. 5.9) to create a better form for transplantation. Right now we are learning the basics on which we will be able to build.

Junk DNA

Only about 5% of the 3 billion base pairs of the human genome codes for protein. The remaining 95% is sometimes called "junk" DNA. This junk DNA is mainly repeating sequences that are derived from a number of different mechanisms. An understanding of these repeating sequences is important when DNA is used for fingerprinting or for studying human anthropology. Repeat sequences also tell us about genetic processes such as mutation.

I will mention only a few features of junk DNA, just enough so that we will remember that it is there, and there in much larger quantities than the DNA that codes for proteins. The repeats that make up junk DNA are sometimes just short simple runs of one or a few bases such as $(A)_n$, $(CA)_n$, or $(CGG)_n$, where n is the length of the repeat, typically 8 to 20. Repeating sequences can also be much longer. Consider the mysterious Alu repeat. Alu is a sequence that is about 280 bp long that is repeated over 1 million times in the human genome! In fact, 10% of the human genome is Alu sequences! The Alu sequences are not evenly distributed throughout the genome; their concentration in certain areas may affect the rate of recombination and gene duplication. Alu and other larger repeat sequences within the genome are increasing in number over time; they are much less common in fossil DNA. Some view repeat sequences such as Alu as a 'parasite' living within our DNA. There is much left to learn about "junk" DNA.

Gene Maps

If you thought that those fan-folded road maps that get all messed up on long car trips are a problem, wait until you try your first trip to a location on the human genome map. The human genome is immensely large and incredibly detailed. Let's go to a specific location as an example—the site of the very important $p53$ tumor suppressor gene. Figure 1.7 displays both a real genome map and as an analogy a geographic map to show scale. The human genome is 3,000 megabase pairs (Mb) long. Two copies of the genome, each slightly different, are contained within the nucleus of every cell. For scale, let's equate this to a map of the earth with a circumference of 40,000 km. Searching the Internet we find that the human $p53$ gene is OMIM (Online Mendelian Inheritance in Man) number 191170 or in the LocusLink database number 7157.[2] These are two of several Internet sources for the human genome map. Table 1.4 lists many of the major gene databases on the Internet. I will often give an official reference number to one of the databases when mentioning a gene. Visit these Web pages and see the information about $p53$ and the multiple links to more information. Remember, the human genome lives on the Internet. All printed sources are outdated facsimiles of the evolving truer picture maintained on the Internet.

The $p53$ gene is localized to a region on the short arm of chromosome 17 (17p13.1). There are 96 Mb of the human genome on chromosome 17; this is equivalent to 1,200 km on our geographic map,

---

[2] *http://www.ncbi.nlm.nih.gov/htbin/Omim/dispmin?191170* and *.ncbi.nlm.nih.gov/LocusLink/LocRpt.cgi?1=7157.*

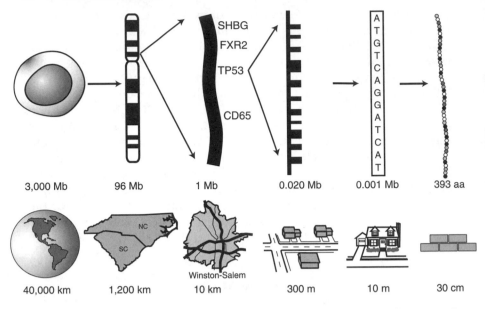

| 3,000 Mb | 96 Mb | 1 Mb | 0.020 Mb | 0.001 Mb | 393 aa |

| 40,000 km | 1,200 km | 10 km | 300 m | 10 m | 30 cm |

Figure 1.7. A map of the human genome is compared to a geographic map of the world to give a sense of scale to structures within the genome.

enough to show North and South Carolina. More precisely, *p53* is located 9.1 Mb proximal to the end of the short arm of 17. A region of 1 Mb of the genome shows the *p53* gene (denoted *TP53*) and several flanking genes such as *SHBG* (sex hormone binding globulin), *FXR2* (fragile X mental retardation), and *CD68* (CD 68 antigen). This span of 1 Mb on our geographic map is equivalent to 10 km, enough to show central Winston-Salem, NC, the town where I live.

Physical distances on the genome are denoted as base pairs (bp) from a specific signpost. Sometimes, as in this case, the end of a chromosome is used. Other signposts are site tags. *TP53* is close to marker D17S1678. Besides measuring distances in base pairs, it is sometimes useful to measure distances in centimorgans. As you will see later in this chapter, genetic distances are a measure of how far genes are apart in terms of their probability of separating at meiosis when the egg and the sperm are formed. This distance is very important to geneticists when describing inherited traits. The activity of recombination at meiosis is about double in the formation of an egg versus a sperm. Thus *TP53* is between 9.8 (male) and 18.1 (female) centimorgans from the end of the short arm of chromosome 17 depending on which sex you are mapping!

The *p53* gene consists of 11 exons that span 20,303 bp. A genomic map with this detail is equivalent to 300 m on our geographic map, enough to show a few houses. When the *p53* gene is transcribed into mRNA, a sequence of 1,179 nucleotides is produced, which translates into 393 aa. The mRNA piece is equivalent to 10 m on our map. Going the last step

Table 1.4. Gene databases on the Internet.

Web address(www)	Database
celera.com	Celera Genomics human genome database
ensembl.org/genome/central	International Human Genome Sequencing Consortium
ncbi.nlm.nih.gov/genome/guide/human	National Center for Biotechnology GenBank
ncbi.nlm.nih.gov/htbin/Omim	On-line Mendelian Inheritance in Man (OMIM)
ncbi.nlm.nih.gov/LocusLink	LocusLink
nhgri.nih.gov/educationkit	National Human Genome Research Institute—educational
ornl.gov/hgmis	Department of Energy genome resources

and specifying the exact nitty-gritty amino acid sequence of the *p53* protein is down to a detail equal to 30 cm. This is the size of one brick.

Mapping the human genome down to the level of the base pair sequence for the p53 gene is equivalent to having a map of our planet that shows where every brick in every building on the planet is located! We know more detail about our genomic code than we know about the surface of our planet.

Writing down the human genome is not the end. We desire to understand the genome functionally. The *p53* gene works in a very complex way, interacting with other genes. Its role is to slow or stop the proliferation of cells with damage to their DNA. Rarely, individuals are born with a defect in the *p53* gene. The inheritance of a mutation in one of the two *p53* genes is called the Li-Fraumeni syndrome and results in early onset of multiple malignant tumors. The majority of cancers, however, are associated with a spontaneous acquired mutation of the *p53* gene within the tumor cells. There are hundreds of different mutations found, but they are grouped into four hot spots in the genome. These hot spots code for portions of the p53 protein crucial to its function. Tumor cells with defective *p53* genes can be targeted by specific molecular therapies such as antibodies or antisense oligonucleotides (as we will see later in Chapter 7). The human genome is more than a catalog or map. It contains knowledge that allows for entirely new forms of diagnosis and treatment.

Physical Organization of the Genome

The picture of the genome that I have given so far is a functional one. Figure 1.4 is a schematic diagram of some of the operational elements of a gene. Even Fig. 1.5, which lists the DNA sequence of an exon, is not at all a physical picture. If we were to look at a piece of DNA, these functional structures would not be visible.

The physical organization of the genome, though very different from the functional anatomy, is also important. The nucleus of a cell is only a couple of millionths of a meter across, yet we have to put 2m of a linear DNA molecule into it. This is a lot of "string" to pack into a sphere. A scaled-up model, making the cell nucleus the size of a basketball, would in turn scale the DNA to the thickness of a spiderweb strand 0.2 mm thick with an overall length of 200 km. The packing of 200 km of spiderweb into a basketball is difficult to imagine. The problem is even more complex when we include the requirement that we must be able to locate any one of the tens of thousands of genes strung out along that strand within fractions of a second (the time it takes for a cell to activate a gene). We also have to unwrap all 200 km as the cell divides. Remember, we need to make two copies of the string over 12 hours and give one copy to each daughter cell.

The physical organization of DNA is accomplished through a successive hierarchy of packing structures as represented in Fig. 1.8. The linear double-helix DNA molecule is wrapped for two turns around small globular proteins called histones. These histone beads are grouped into a winding helical secondary structure of their own. Then the secondary structure is packed into loops. When a cell undergoes mitosis, these loops gather together in a structure that makes up the arms of the visible chromosome. There are 46 chromosomes in every human cell. These consist of pairs of chromosomes 1 to 22, plus the sex chromosomes XX in a female and XY in a male. Within each pair, one chromosome comes from the mother and one from the father.

A picture of even a very small portion of DNA would appear as an incredible tapestry woven of a long, thin molecule wrapped around the histones, like thread on bobbins. So there is really a tremendous hierarchy of physical structure to DNA, with at least four levels of increasingly complex structure. We do not understand at all how a small portion of the genome can be accessed quickly. How can a single piece of DNA be located, opened, read, and closed? It is important to realize what we do not know as well as what we do.

Basic Genetics

Genetics is a large subject and one that has been completely rewritten by the molecular revolution. No longer do we need to breed fruit flies (*Drosophila*) in the laboratory. Innumerable crosses between individual organisms to probe their genes are made obsolete by our ability to probe DNA directly. The great barrier to human genetics has been removed. In the past we had been limited to observing pedigrees of a few families with phenotypic abnormalities. Now we can probe the human genome directly. In Chapter 5 we will see that we can determine by a

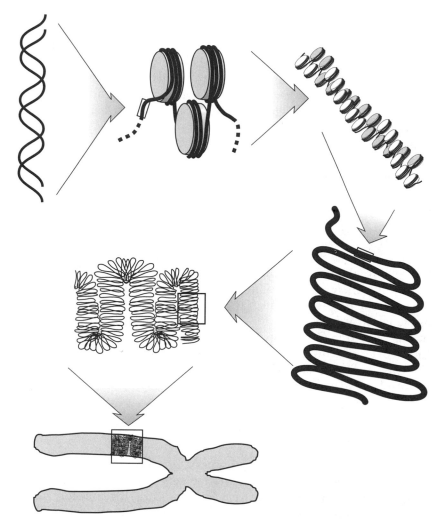

Figure 1.8. The physical structure of the genome shows the multiple levels of packing DNA into its final chromosome structure.

laboratory test whether a patient is a heterozygote for the hemochromatosis gene. We do not need to ask whether his Uncle Harry had the disease in order to guess at the statistical chance that he is a carrier. Our review of basic genetics here will be succinct, because the examples in the rest of the book will demonstrate the principles.

Alleles and Inheritance

The human genome is diploid, with two copies of every chromosome. However, the chromosomes are not necessarily identical. Each gene has a specific location on a chromosome, the locus. At a given locus there is one copy of multiple alternative forms of a gene. Consider the gene for the beta chain of hemoglobin located on chromosome 11. There are many alleles of this gene. In fact, hundreds of variations of the normal hemoglobin chain genes are known. One diploid human will possess two alleles of the beta chain gene, one on each of the two chromosomes 11. Many of the allelic variations of the beta-chain gene produce no disease or detectable abnormality. Some, like the sickle cell allele (*Hgb S*), produce a laboratory abnormality (in vitro sickling) when present as one copy, and a clinical disease when present as two copies. The sickle cell gene is **recessive**. This means that if only one copy of the abnormal gene is present, disease is not apparent since the remaining normal beta-globin allele is capable of producing sufficient functional protein. If only one allele present is sufficient to cause an abnormality, the allele is said to be **dominant**. The *Hgb S* gene differs from the normal *Hgb A* allele by only a base, a T for A substitution in codon number 6, base number 20. *Hgb C,* another variant hemoglobin that crystallizes rather than sickles, has an A for G substitution at base 19, as shown in Fig. 1.9. We saw earlier in Fig. 1.6 that there are two common alleles for the proinsulin *INS* gene, with four base substitutions.

Polygenic Traits

Starting with the simple and progressing to the more complex is usually a good way of learning. Mendelian genetics, one gene–one trait, is simple but misleading in that very few traits are this simple. Up to this point in human genetics, we could handle only simple traits; all of the examples in our older texts of "genetic" diseases discuss simple traits. The reality is that most traits or phenotypes are dependent on complex interactions between many genes. Diabetes and atherosclerosis are examples of polygenic traits. These will be discussed in detail in Chapter 5. The immune system is even more complex; its behavior is dependent on interactions of many genes and events that happen to the organism later. We discuss this in Chapter 6.

We are at a new level of sophistication in genetics, very far postmendelian. We will see that the terms *recessive* and *dominant* apply only in a few situations. For polygenic traits and human disease, we must be ready to expect phrases like the following: "Inheritance of the homozygous E4/E4 allele of apoE lipoprotein is associated with increased risk of Alzheimer's dementia and early-onset atherosclerosis." E4/E4 does not cause either of these diseases and its inheritance does

Codon no.		5	6	7
Base no.			20	
Hemoglobin	A	CCT	GAG	GAG
		Pro	Glu	Glu
Hemoglobin	S	CCT	G**T**G	GAG
		Pro	Val	Glu
Hemoglobin	C	CCT	**A**AG	GAG
		Pro	Lys	Glu

Figure 1.9. The genetic sequences for a portion of the beta chain of hemoglobin are shown for the normal hemoglobin A allele and for the mutant S and C alleles. Both of these abnormal hemoglobins are the result of a single base error.

not mean that an individual will certainly suffer these problems. E4/E4 tips the balance in a complex biology of lipid metabolism somewhat in the direction of atherosclerosis. Genes are the building blocks of the body, but the complex structures that result depend on much more than the name of every block.

Human Genome Project

The Human Genome Project, a plan to bring "big-science" resources to the challenge of sequencing human DNA, began in 1990 and effectively ended in February 2001. The end, which is really just a new beginning, was the publication of a draft of the human genome in the two major science journals—*Nature* on February 15 and *Science* on February 16, 2001. The draft published in *Nature* was the one produced by the government-sponsored Human Genome Project itself. *Science* published the human genome developed by Celera Genomics, a competing private company that employed a different method for sequencing. In the end, the two cooperated to a good degree and produced very similar results. The best way to access the human genome is through the Internet (Table 1.4).

Let's discuss some of the most important general findings that resulted from sequencing the human genome. First, whose genome is it? Pooled DNA from a small number of people was used as the source, so the human genome is really the average of a small number of individuals. Each of us will deviate from the published human genome because of our uniqueness. To begin, each of us differs at about 1 in every 1,000 bp. There are about 2.5 million **single-nucleotide polymorphisms (SNPs)** in the genome. These are one base differences that characterize our individual variations in noncoding DNA.

How many genes are there? In Table 1.1, I list the number of genes at 35,000. The final number is not yet in, but it will likely be close to this. A few years ago, estimates of the number of human genes would have been close to 100,000. The low number surprises us. However, human genes are showing themselves to be more versatile than their numbers suggest. Alternate splicing of exons allows a single gene to code for more than one protein; three proteins per gene may be more typical.

What do we do with the information of the human genome? The next steps are understanding what we have written down. The "hot-button" words now are *proteomics* and functional **genomics**. Having the human genome database greatly speeds up all kinds of research. Every time we find a growth factor, a signal molecule, or a transmitter; we can look into the genome and find where it came from. Similarly, we can look in the genome for sequences that code proteins with the characteristics of a neuropeptide or a transmembrane receptor. We find sequences that look promising because they resemble genes that we already know. These sequences then lead us to the new gene and its protein product. The human genome is a library of information that is now open for business. There is a treasure of information in the human genome leading to new drugs, new diagnostics, and new medical knowledge. The next phase of discovery is learning to use the library that we have so successfully copied from our own DNA.

Aside: What the Genome Does Not Code For

It is useful—and sometimes reassuring—to consider what the human genome does not specify. Ironically, one thing is fingerprints. Identical twins have identical DNA, but they do not have identical fingerprints. Fingerprints develop; they are not genetically determined. The immune system is more than the result of a genetic blueprint. It begins with proteins coded by genes, but grows into something larger, not specified in the genes. In response to a history of chance exposure to various antigens, the system becomes unique in every person. After bone marrow transplantation it reconstitutes itself, and with a new history of antigen exposure it becomes a system different from the prior immune systems of either donor or host.

DNA does not encode the functioning of the human brain. The proteins and the basic building plan are coded in the DNA. However, the brain's neurons have vastly more interconnections than can possibly be specified even if all 3 billion bp in the genome were devoted to that one subject. The interconnections and hence the function of the brain are built first as a self-organizing system and second by experience. The self-organization of the early central nervous system is something marvelous and beyond our current understanding. If a dozen or so primitive nerve cells are placed in a culture dish, they will start to form interconnections. These tentative connections will re-form several times, exchanging electrical nerve impulses. After some days, the nerve cells will have formed a network; the connections are finalized. The entire brain appears to build itself by self-directed organization in similar fashion.

After the brain is built it begins to learn, acquiring what we call knowledge by experience, instruction, and conscious thinking. That knowledge does not flow back into the genes, as was thought by some early biologists. We do not inherit acquired knowledge or behavior. Humans and some animals can pass their knowledge to their offspring, but it is by teaching, not genes.

This may change! If we write into the genome of humans and animals, as we are about to do with genetic engineering, then the genome does become the product of acquired knowledge. A **transgenic** cloned human poses a terrible problem of what constitutes a person. We are much more than our DNA, because our DNA alone is neither alive nor capable of specifying all that we are.

Summary

This chapter has laid out the basic structure of the human genome; blown up to a million miles of railroad track or 200 km of spiderweb in a basketball. The vast information content of the genome has been compared to a library and to a map of the world. We must strive to appreciate the scale of the genome. Dissecting the problem down, we considered the anatomy of a few genes, BRCA1, C-MYC, and INS, all of which will reappear later in the book as clinical examples. Basic genetics is changed by our molecular knowledge; inheritance is not as simple as Mendel proposed. One gene may code for many proteins, and many genes may be involved with complex diseases such as diabetes and atherosclerosis. The Human Genome Project has been completed; we are left with a wealth of information that the rest of this book will explore.

Bibliography

Alberts B, Johnson A, Lewis J, Raff M, Roberts K, Walter P. *Molecular Biology of the Cell*, 4th ed., New York: Garland, 2002.

The human genome issue. *Nature* 2001; 409:813–958.

The human genome issue. *Science* 2001; 291:1177–1351.

Meldrum DR. Sequencing genomes and beyond. *Science* 2001;292:515–517.

Ridley M. *Genome: Autobiography of a Species in 23 Chapters*. New York: HarperCollins, 1999.

2   Gene Expression and Regulation

Overview

The unique feature of DNA is that it is duplicated at cell division, with one copy going to each daughter cell. The genomic information that makes us what we are is thus faithfully passed into each new cell. The process of DNA replication is coupled with DNA repair systems to make certain that errors in copying are corrected, or that the copy is destroyed. Mitosis is the overall act of cell division involving sorting out of the duplicated DNA on chromosomes into each daughter. Meiosis is a very special form of mitosis used to produce haploid germ cells. The regulation of gene expression shows that DNA not only carries information; DNA also contains regulatory elements that control the flow of information. Mutations are errors in DNA; some are bad, many are the source of our ability to evolve to a changing environment.

DNA Replication

The replication of DNA begins with a separation of the double-stranded DNA helix. This process is called "melting" because it takes the equivalent of thermal energy to occur. Heating double-stranded DNA in vitro to 90°C will cause it to separate into single strands. In the cell, DNA strand separation begins at multiple points throughout the genome. Enzymes catalyze the process of strand separation, allowing the DNA to melt at 37°C. Each local area where strand separation occurs is the start of a structure called a replicon. Several enzymes are involved in initiating and controlling this process. An enzyme called DNA helicase controls the unwinding of the double-stranded DNA molecule. Another enzyme, DNA polymerase, catalyzes the synthesis of a new duplicate strands using the unraveled original DNA strands as

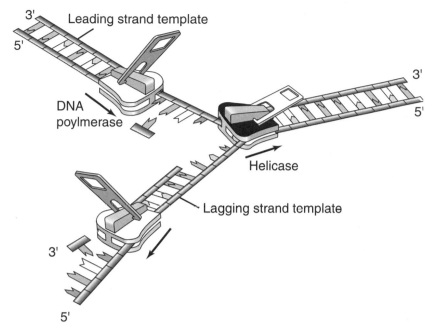

3'

5'

Leading strand template

DNA
poylmerase

3'

5'

Helicase

Lagging strand template

3'

5'

Figure 2.1. The process of DNA replication unwinds the double strands and then uses each strand as a template for a new copy.

templates. Figure 2.1 demonstrates this process by showing a single DNA replicon. As the DNA replication extends in both directions, the newly synthesized strands re-form a double helix with each of the original template strands. This is called semiconservative replication. Each copy of the DNA will contain one original strand from the parent molecule and one newly synthesized strand twisted together in a double helix.

Mitosis and Meiosis

Whenever a cell divides, it must give exact copies of its genome to each daughter. Most cells are in a resting state called G0. Appropriate signals can tell the cell that its useful life is over, and the cell then undergoes programmed cell death, called **apoptosis**. Other signals can call the cell back into active replication from G0. A cell preparing for replication enters the cell cycle at a point called G1 as diagrammed in Fig. 2.2. An initial step is to check the integrity of the DNA using several repair mechanisms, as we

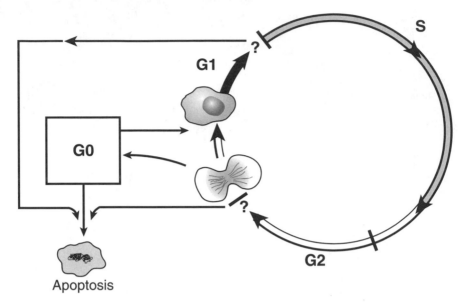

Figure 2.2. The cell division cycle is defined by the major events that control proliferation. Actively dividing cells progress from G1, to DNA synthesis (S), to G2, and then to mitosis (M). At several checkpoints, denoted by ?, the cell's progress is tested and if damage is detected, the cell dies (apoptosis). G0 is a reservoir of quiescent, nondividing cells.

will discuss later in this chapter. If the cell passes muster, the entire DNA content is duplicated over a period of approximately 12 hours (S phase). At the end of S phase, a cell has twice diploid (tetraploid) DNA. The chromosomes have been doubled from 46 to 92. A short pause follows (G2), and then the cell begins mitosis. At mitosis, the DNA condenses into visible chromosomes. At all other times the chromosomes are in a more dispersed configuration. In mid-mitosis, the chromosomes line up, with each pair of chromosomes together. The cell then begins to divide, with half of the chromosomes migrating to each of the forming daughter cells. Each daughter cell gets an identical copy of the genome that includes a paternal and maternal copy of each chromosome.

Meiosis is a more complex process that occurs only in the formation of the germ cells, the egg and sperm. Germs cells are haploid, possessing only one copy of each chromosome. At conception when the egg and sperm fuse, the normal diploid genome is restored. To produce haploid germ cells, the parent cell after doubling its DNA must undergo cell division twice. At the first division, all 92 chromosomes are lined up in the center of the cell with the four copies of each chromosome close together. The genetic mechanism of recombination occurs at this time. The arms of homologous chromosomes are **crossing over** each other and the ends are swapped. Imagine four strands of spaghetti loosely organized parallel to each other. As pieces flop over each other, they are cut and spliced to switch ends. Recombination allows for the formation of new chromosomes that are a mixture of parental and maternal strands from the previous generation.

Longer chromosome strands have a greater chance of crossing over. One way of expressing the length of a chromosome is the distance based on the probability of crossing over. This distance is a statistical estimation of the chance that genes on the same chromosome will separate in future generations. The centimorgan is the unit of genetic distance and implies a 1 chance in 100 of crossover. Roughly a centimorgan is equivalent to 1 megabase in physical distance. As I mentioned in Chapter 1, recombination in women occurs twice as frequently as in men. There are also numerous hot spots along the chromosomes where crossing over occurs much more frequently. So distances measured in centimorgans are not exactly linear. Recombination via crossing over is a very important genetic mechanism. The four germs cells produced during the second meiotic division have far greater genetic diversity than just representing a faithful copy of either a maternal or paternal chromosome of the previous generation. The genes are re-sorted at every generation; only those that are close together stay together.

DNA Repair

When a dividing cell replicates its DNA during the S phase of the cell cycle, a number of errors will be made. Our best estimate is that about 600,000 (0.01%) of the 6 billion base pairs will have been copied incorrectly during the first pass of DNA synthesis. The DNA synthesis mechanism is very, very good but not perfect. Fortunately, most of the errors made during DNA synthesis will be corrected. If they cannot be corrected, the cell will in all likelihood be destroyed later at a checkpoint. There are a number of DNA repair mechanisms operating within the cell nucleus that will improve the error rate from the 10^{-4} seen just after DNA replication to approximately 10^{-9} before the cell division cycle is complete. That means that out of 6 billion base pairs only one error will remain! DNA repair works not only in correcting copy errors, but also in correcting other damage to DNA that occurs in nondividing cells.

Mutations

If damage to DNA is unrepaired and results in a change in the genome, then we have the potential for a mutation. Not every alteration in DNA produces an abnormality; in fact, most do not. If a change in DNA occurs between genes or within an intron, generally no effect occurs. Such a change is called a polymorphism. A difficult distinction in the definition of a mutation is whether a given change is an abnormality or just different. Some people use the word *mutation* only for a deleterious change in a gene, and call all other changes a polymorphism. By this criterion, the two alleles of *INS*, alpha and beta, are polymorphisms since no known effect is produced by alpha versus beta. Recall from Table 1.1 that there are about 2.5 million single-nucleotide polymorphisms (SNPs) in the human genome. These are differences, not mutations. Another example that tests the definition of a mutation is baldness, a phenotype that describes me. Is this an acceptable variation on normal hair bearing, or an abnormality? Whatever your answer, I must tell you that almost all geneticists would respond that I carry the mutant gene for male-pattern baldness.[1]

Another important distinction for us to make is between an acquired **somatic-cell** mutation and an inherited **germline** mutation. Classical genetics deals mostly with inherited traits due to genes present in the germline. Somatic mutations arise spontaneously and begin in a single cell as the result of an error in DNA copying or repair. Most cancers are due to somatic mutations resulting from DNA damage.

Point Mutation

Mutations come in many sizes and kinds; I have listed the most important with a few examples in Table 2.1. Sickle cell anemia is due to a point mutation in the beta chain of hemoglobin. The mutation is a T replacing A at the 20th position. This results in a change in the 6th codon that causes valine to replace lysine (recall Fig. 1.9). *Hgb S* is considered a mutation because it produces a serious life-threatening illness. This occurs, however, only when the mutated form has replaced both copies of the gene; this is a recessive gene. Sickle cell trait occurs when only one copy of the gene is mutated. Sickle cell trait does not produce an illness, and in fact may offer some protection against malaria.

Frame-Shift Mutation

BRCA1 is called the breast cancer susceptibility gene. It belongs to a class of genes known as tumor suppressor genes that function to repair DNA damage and errors of cell division that, if uncorrected, lead to tumors. As we will see in Chapter 7, most cancers do not spread until

---

[1] I also shave my head. This is confusing to a geneticist. Am I really bald by inheritance, or is my phenotype an acquired act that will not be inherited?

Table 2.1. Types of DNA mutations.

Name	Example
Point mutation	T for A in base no. 20 hemoglobin S
Frame shift	del185AG in *BRCA1* breast cancer susceptibility
Chromosomal translocation	t(9;22) "Philadelphia" in chronic myeloid leukemia

they have broken free of the supervision provided by the tumor suppressor genes. Most mutations in tumor suppressor genes are acquired somatic mutations that occur only in the tumor tissue; only a few mutations in these genes are inherited. *BRCA1* is known to have hundreds of different mutations and is the source of much of the inherited risk of breast cancer. Any mutation that disrupts the protein of *BRCA1* increases the risk of breast and ovarian cancer in mid-life. One of the most common mutations in *BRCA1* is 185delAG, a frame-shift mutation. In 185delAG, the bases A and G at position 185 and 186 are deleted. Recall our experience with open reading frames in Chapter 1, and guess what happens. The entire remaining message after position 185 is read out of sequence; the open reading frame of a normal gene is lost. Figure 2.3 shows what happens. The normal mRNA sequence is shown above, the frame-shift mutation below. After A and G are removed, all of the following translated amino acids are changed. After 17 codons producing random amino acid translation, the stop codon TGA occurs and protein synthesis ceases.

Chromosomal Translocation

The processes of mitosis and meiosis are complicated. Sometimes when DNA is damaged or broken, chromosomes will be assembled incorrectly. A chromosomal translocation results from the broken ends of two different chromosomes being exchanged. The genes near the break points are disrupted. Chromosomal translocations are almost always acquired somatic mutations. The Philadelphia chromosome t(9;22) occurs only in the malignant white blood cells of chronic myeloid leukemia (CML). We now know one gene on chromosome 9 (*ABL*) becomes inserted into the middle of another gene on chromosome 22 (*BCR*). The deregulated *ABL/BCR* fusion gene is a mutation that (as we will see in Chapter 7) leads to abnormal cell proliferation in white blood cell precursors. In Chapter 8, we will see that this knowledge of the *ABL/BCR* mutation allows us to tailor-make a drug that blocks the abnormal protein produced by this gene fusion. This drug, STI571, is a new very effective treatment for CML.

Normal

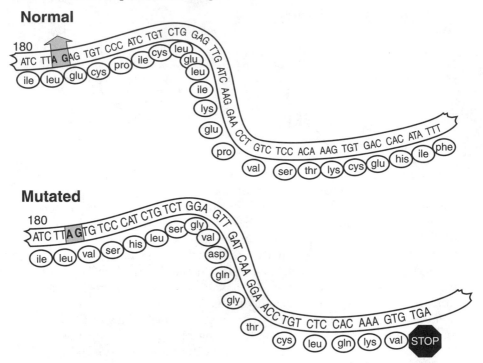

Mutated

Figure 2.3. The mRNA sequences for bases 18 to 248 of *BRCA1* for the normal and a mutated form are shown. The del185AG mutation causes a frame shift that results in the early termination of the protein with a stop codon, TGA.

There are many other kinds of mutations, such as deletions of large segments of a chromosome or reduplication with extra copies of genes. We have a lot to learn about mutations. Although the very word *mutation* conjures up visions of horrible abnormality, mutation is a mechanism by which organisms adapt to their changing environment. Genetic engineering constructs altered genes for beneficial purposes. Would we consider these changes mutations? As our manipulation of the genome progresses, our definition of mutation will itself have to evolve.

Aside: Telomeres and Aging

Aging is the decreased repair of damage to tissues and loss of function that occurs over an organism's natural life span. Aging is dramatically species specific. The natural maximal life span for a human is about 120 years, 40 for a chimpanzee, 30 for a dog, and only 3 for the laboratory white mouse, yet all age similarly. A human infant and a puppy may start life together, but by the time the child is a teenager, the dog will be old with stiff joints and sagging, wrinkled skin.

Most theories of aging center around the accumulation of unrepaired cellular and molecular damage. There also appears to be some limit of the total number of cell divisions that are normally allowed to occur. In a culture dish, cells from older people will undergo fewer divisions than those from children. After about 50 divisions, cells become senescent. Fusion of a senescent cell with a young cell inhibits division of the young cell. Addition of mRNA from old cells has the same effect.

Multiple pathways appear to trigger cellular senescence. One such pathway is **telomere** shortening. Telomeres are genetic elements located at the distal end of chromosomes. Telomeres consist of about 15,000 bp of repetitive DNA sequences. The six bases TTAGGG are repeated thousands of times. These repetitive sequences are replicated by a special DNA telomerase. However, at each division, a small piece of the telomere, about 100 bp, is not copied. Thus, telomeres become shortened over progressive cell divisions. At some point, this shortening induces inhibition of DNA replication and cellular senescence. Cancer cells rebuild the cleaved piece from the telomere at each division. Cancer cells are immortalized as part of their abnormal function, independent of the normal countdown of telomere shortening.

No one thinks that aging is due to just one gene or one process. Telomere shortening is one very interesting part of the problem. With genetic engineering we can alter cells to rebuild their telomeres; will this alter aging?

DNA to Protein

When a gene is turned on, the portion of DNA that encodes the gene is transcribed from DNA to mRNA. The intron portions are spliced out of the message. The edited mRNA is than translated from RNA to protein on the polyribosomes of the cell's cytoplasm. A chain of amino acids representing the primary structure of the final protein is produced (recall Fig. 1.3).

After translation, the protein undergoes posttranslational modification. These changes include cleaving the primary chain or joining several chains, adding nonprotein groups such as sugars, and folding of the protein into a complex three-dimensional structure. The posttranslational changes are not specified by the genome. They happen when the primary form of the protein occurs in the correct cellular environment.

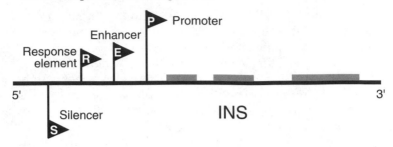

Figure 2.4. The *INS* gene is regulated by a number of control elements that are upstream of the coding sequences.

Proteomics

Already with the Human Genome Project having achieved mapping of the DNA, a second database is being built. What forms do the proteins specified by the genome take? This listing is called **proteomics**. The catalog of all of the cell's proteins will be more difficult to assemble than the genome. One gene can produce several proteins. We cannot predict what functional form a complex protein will take, even if we know its genetic code. Look back at our *INS* gene map (Fig. 1.6). We know the genetic sequences and the structure of the gene. This is a simple gene; only a small amount of posttranslational modification occurs—two snips and one paste. Yet, we could not have predicted the complete form of the insulin protein from the gene. Proteomics will be the next wave of discovery as we learn how genes are turned into proteins.

Gene Expression

Insulin

The control of gene expression is complex. Consider the problem using the insulin gene as an example. We know that insulin is produced in the pancreas (and not anywhere else) in response to the body's need. There must be signals that tell the insulin gene when to turn on and when to turn off. These signals must be keyed on stimuli like glucose level, stress, type of food intake, and many other factors.

We do not know all the details, but let us consider some basics of the regulation of insulin gene expression. This will show the problem nicely. There is a core **promoter** region just upstream (about 100 bp) of the insulin gene. Actually there are several promoters. **Transcription factors** are proteins that bind to DNA. In general, gene expression is initiated by the binding of transcription factors to the promoter. Thus,

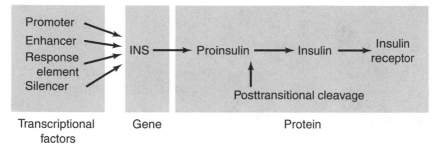

Figure 2.5. The regulation of insulin continues with posttranslational and other biological steps, beyond what is coded for in the gene.

a transcription factor that binds to the insulin promoter initiates transcription of the *INS* gene into mRNA. Several transcription factors for the initiation of the *INS* gene are tissue specific. They are present only in the beta islet cells of the pancreas. This is another important feature of gene expression; it must exert control of events both in time and in location. We want insulin made in the pancreas and hemoglobin made in the red blood cell, and only when we need them.

Just turning a gene on and off with a transcription factor would work, but fine control and interaction with other events is desirable. Thus, gene expression is further regulated by many other factors beyond a transcription factor binding to the promoter. We can kick up the level of initiation with an enhancer. Or slow it down with a silencer. Or change the level based on some specific event with a response element. These additional regulators are portions of DNA that are sometimes close and sometimes far from the gene they help. They add their control when specific proteins bind to these regions. Figure 2.4 shows an expanded map of the *INS* gene with some of these regulatory elements indicated. In Chapter 5, we will look at a genetically engineered version of the *INS* gene, for purposes of transplanting insulin function into diabetic patients. We will have to keep in mind that in addition to transplanting the gene, we will need to add regulatory sequences to our engineered gene to make it function.

Now we are getting somewhere. We can regulate the insulin gene with promoters, enhancers, silencers, and response elements. You can see how the system builds to the rich detail that we imagined would be necessary. It gets even better. We can affect the production of insulin posttranslation. Recall that the *INS* gene produces a precursor peptide that must be cleaved into two chains that join side by side. We can influence how fast this happens with additional regulatory events. Furthermore, the body's response to insulin is also dependent on the pres-

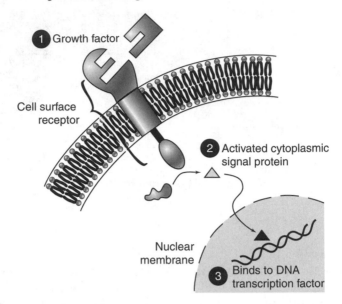

Figure 2.6. The cell signaling pathway receives external signals that are carried stepwise into the nucleus. The end point is an interaction with DNA that turns on a gene via its promoter.

ence of insulin receptors on cells in muscle and other target tissues. The equally complex control of expression of the insulin receptor gene likewise influences that end of the system. Figure 2.5 schematically demonstrates many of the levels of control of gene expression related to insulin. Insulin is one of the smallest and simplest genes in the body. Even though this looks complex, this is about as simple as gene regulation gets!

Why does all this detail concern us? This book is on molecular medicine, not basic molecular biology. The new treatments and possibly cure of diabetes mellitus will be derived from a knowledge of the information in Figs. 2.4 and 2.5. We have heard talk of pancreatic islet cell transplantation. What if the problem is in the receptor to insulin rather than in its production? Why transplant islet cells, when the *INS* gene is in every cell in the body anyway? Why not just change transcription factors or other control elements to override the defect in diabetes? We will discuss this in Chapter 5. You can skip to that now if you like. These basic concepts are important and immediately applicable to current medical problems.

Cell Signaling Pathway

The signals that regulate genes come from outside the cell. This information must somehow be transmitted across the cell membrane down into the nucleus. The cell signaling pathway is the means for doing this. Figure 2.6 schematically demonstrates this stepwise progress of information. A growth factor present as a soluble molecule in the blood or as a molecule fixed to the surface of a cell is the start of the signaling pathway. The growth factor binds to its specific receptor on the surface of the target cell. Insulin binding to an insulin receptor on a muscle cell is an example. The binding of the growth factor to its receptor changes the shape of the receptor, affecting its intracytoplasmic piece. This shape change carries the signal generated by the growth factor across the cell membrane without the growth factor entering the cell. The signal is further propagated through the cytoplasm to the nucleus by one or more intermediate molecules such as the G proteins. The signal finally reaches the nucleus as a specific piece of protein (transcription factor) binding to a promoter (or enhancer, silencer, response element, etc.) on the DNA.

In Chapter 7, we will see that the cell signal pathway is important in the molecular causes of cancer. The proteins of the cell signal pathway are coded for by oncogenes. When these genes are mutated, the pathway malfunctions. If this malfunction results in unregulated cell growth, we have the first step toward a malignant tumor.

Summary

DNA replication reproduces the information of the genome. Transcription of DNA into RNA and translation of RNA into protein are the expression of genomic information. These processes are regulated and subject to repair systems that ensure the correctness of the information. Aging of cells is in part controlled by telomere shortening, which eventually limits the number of times a cell may divide. Mutations are errors in information, sometimes inherited within germ cells and sometimes acquired in somatic cells. Mutations are the cause of many diseases and also the source of new genetic diversity that permits adaptation to change. Outside events trigger gene expression via the cell signal pathway. These signals bind to control elements on DNA, making the cell interact with other cells, the whole becoming a multicellular organism with greatly increased complexity of function.

Bibliography

Lewin B. *Genes VII.* Oxford, UK: Oxford University Press, 1999. (This book is also available on-line at *www.ergito.com.*)

Lodish H, Berk A, Zipursky SL, Matsuraira P. *Molecular Cell Biology* 4th ed. Philadelphia: WH Freeman, 1999.

Maroni G. *Molecular and Genetic Analysis of Human Traits.* Malden, MA: Blackwell Science, 2001.

Strachan T, Read AP. *Human Molecular Genetics*, 2nd ed. New York: Wiley-Liss, 1999.

3   Genetic Engineering

Overview

Molecular medicine is a revolutionary new field that exists because of the creation of new tools for examining and manipulating DNA. The invention of the microscope showed us the cellular basis of life. The invention of recombinant DNA technology has now revealed the molecules of life. The revolution is only 25 years old and so much has already been learned. We will begin this chapter with an explanation of the first tool, a DNA scissors made possible by restriction enzymes. The myriad of techniques that followed—Southern blot, polymerase chain reaction (PCR) and DNA on a chip—all resulted from learning how to hybridize small probes to molecular targets. These methods then led to further ways to manipulate the genome, shutting down individual genes with antisense or the knockout mouse model. The latest step in genetic engineering is the full capability to clone; copying and modifying genes or even entire organisms.

Restriction Enzymes

Restriction enzymes are bacterial proteins that cut DNA into fragments. A restriction enzyme recognizes a specific base sequence such as AGCT, and cuts DNA whenever that combination of letters occurs. This place on the DNA is called a restriction site. It seems peculiar that bacteria would produce such an enzyme. However, the function of these enzymes is to destroy bacteriophages or viruses that invade bacteria. Bacteria presumably gained this enzyme function as they evolved, producing a primitive immune system for themselves. Restriction enzymes in bacteria only cut foreign DNA, not bacterial DNA. Bacteria only occasionally have the specific base sequence that would be cut by their own restriction enzyme. Furthermore, bacteria protect their own restriction sites by chemically modifying the DNA with a methyl group rendering it resistant to the enzyme.

The human genome (6,000 Mb) is so much larger than a bacterial genome (1 Mb) that an immune system based on restriction enzymes would not work for us. There are too many restriction sites in our DNA. However, it is just this property that makes restriction enzymes such a useful tool. The EcoRI (pronounced "echo R 1") enzyme cuts at GAATTC and produces tens of thousands of breaks in human DNA. BglII (bagel 2) recognizes AGATCT. The rare cutter NotI (not I) cuts at GCGGCCGC. This is an unusual sequence in human DNA. There are only a few NotI restriction sites. Altogether, there are hundreds of restriction enzymes, providing a very flexible system for engineering cuts in DNA.

The use of restriction enzymes as a tool in DNA technology can best be demonstrated with an analogy. The paragraph below contains a copy of the preamble to the Constitution of the United States in which there is an error. We will use the technique of restriction enzymes to find the error. I bet that you cannot find it by yourself. Imagine a restriction enzyme that recognizes the sequence "the." Find within the preamble every occurrence of the sequence of letters "the" and make a mark through the "t." Count the number of letters and spaces between each mark and write down these numbers. These numbers represent the lengths between each restriction cut.

We the People of the United States, in Order to form a more perfect Union, establish Justice, insure domestic Tranquility, provide for the common defense, promote our general Welfare, and secure the Blessings of Liberty to ourselves and our Posterity, do ordain and establish this Constitution for the United States of America.

A friend in Washington, DC, has done the same exercise while looking at the original preamble to the Constitution. He says that his restriction cuts produce fragments of 4, 14, 28, 28, 32, 103, and 118 characters long. This is different from the result you obtained. Can you determine where the mutation in your copy has occurred?[1] This technique of determining mutations is called restriction fragment length polymorphism (RFLP) analysis. In Chapter 5, we will use RFLP analysis to find a point mutation in the coagulation factor V gene. You can jump ahead now if you want to see an actual example of this technique.

---

[1]The error is in the 27th word; "our general Welfare" should be corrected to "the general Welfare." This error could have a significant impact on the interpretation of the Consitution and is a subtle but dangerous mutation! Because of the missing "the" restriction site in the mutant copy, your fragment lengths should have been 4, 14, 28 (28 + 32 =) 60, 103, and 118 characters long.

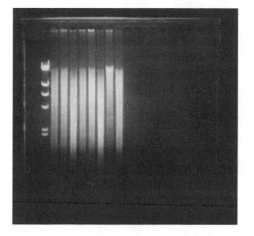

Figure 3.1. An agarose gel electrophoresis of DNA. Lane 1 shows several bands that serve as size markers. Lanes 2 to 9 shows a smear of the thousands of DNA fragments that result from digestion of DNA with a restriction enzyme.

Southern, Northern, and Western Blots

The exercise we have carried out in analyzing the preamble to the Constitution is equivalent to the method called Southern blot analysis. This method detects variation in DNA sequences by cutting DNA with a restriction enzyme and seeing what size fragments are produced. In the actual process of a Southern blot, DNA is extracted from a tissue sample, cut into fragments with a restriction enzyme, and then electrophoresed to separate the fragments according to size. Figure 3.1 shows an agarose gel electrophoresis of DNA. Lane 1 contains size markers consisting of DNA fragments of known length. Lanes 2 through 9 each show a smear of thousands of DNA fragments resulting from cutting the DNA with a restriction enzyme. The final step is to blot this gel onto a membrane and then probe the blot for a specific gene sequence of interest by the technique of DNA hybridization (to be discussed shortly). Figure 3.2 shows a Southern blot of the heavy chain portion of the immunoglobulin gene prepared in my clinical lab. Patients with an extra (rearranged) band have a B-cell lymphoma. This process of analyzing DNA by restriction digest, electrophoresis, blotting, and probe hybridization is named a Southern blot after Edwin Southern, who developed it in 1975. The same procedures carried out on RNA or protein are called a Northern or Western blots, respectively. These names are jokes on the naming of the first blot as Southern. The Southern blot was a landmark technique. Prior to 1975, the only way to probe the genome of an organism was to do breeding experiments!

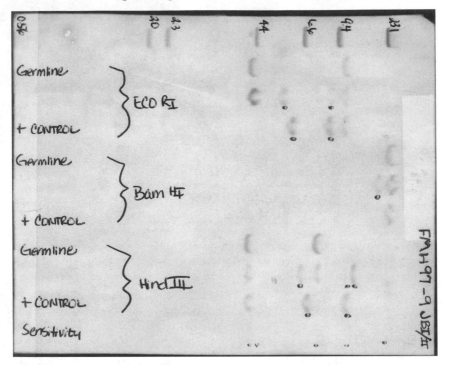

Figure 3.2. A Southern blot analysis of the T-cell receptor gene shows rearranged bands (marked with dots) in several DNA samples from patients with malignant T-cell lymphoma.

DNA Hybridization

Our ability to probe the human genome is based on the affinity for complementary base sequences to recognize each other and to form a reversible chemical bond. To find a specific gene, we make a probe that is complementary. The probe needs to be labeled in some way, with a colored, luminescent, or radioactive tag attached. We then incubate our DNA sample with the probe under conditions that encourage binding of the probe to the target. These conditions include choosing the temperature, pH, and salt concentration. By slightly altering the conditions, we can require an exact match between our probe and target, or just an approximate match. Varying the conditions is called choosing the stringency of the hybridization. DNA hybridization permits us to see our target DNA, as was demonstrated in a Southern blot shown in Fig. 3.2.

DNA hybridization is a key tool for many of the methods of DNA study. As we will shortly see, the famous PCR technique begins with hybridizing a short piece of DNA to a target. DNA on a chip technology also depends on hybridization between our sample DNA and a huge number of probes spaced in an array on a solid substrate.

Polymerase Chain Reaction (PCR)

The **polymerase chain reaction (PCR)** was discovered by Kary Mullis in 1983 and was immediately recognized as immensely useful. The story of its discovery (see Mullis, 1990) is fascinating and emblematic of the biotechnology revolution—young people thinking and acting "outside the box." Mullis was driving from his San Francisco lab to a weekend vacation along the Northern California coast. During the late-night drive, while his girlfriend slept, Mullis pondered a way to make repetitive copies of a single piece of DNA. The series of switchbacks in the narrow coastal road may have entered his subconsciousness. By the time he had arrived at the cabin, he had thought out a chain reaction technique. Within months, he had a working method that rapidly became used around the world. Mullis received the Nobel Prize in 1993. PCR replaced the Southern blot after 10 years, and now, after another 10 years, PCR is being replaced by DNA on a Chip. PCR remains a key technique in many situations, and it demonstrates another magnificent way to manipulate DNA.

I will describe PCR briefly. I warn you that most people need it explained two or three times before they get it. That's not because it is difficult, but because it's so simple. Like a magic trick, it's hard to see! The starting sample of double-stranded DNA is added to a mixture that contains two short single-stranded DNA synthetic probes called amplimers. The amplimers are complementary sequences to a region of interest along the DNA molecule. One amplimer binds on the sense strand and the other binds on the opposite (or antisense) strand, bracketing our target on the sample DNA molecule (Fig. 3.3). The DNA is then heated to near 100°C, which causes our sample to open up and "melt" into two single strands. The reaction mixture is then cooled to about 60°C. At this temperature, the amplimers bind on either side of the target DNA sequence. An enzyme called DNA polymerase begins making new double-stranded DNA. The DNA polymerase requires a short double-stranded region of DNA to prime its copying function. Therefore, new double-stranded DNA is synthesized only at the locations where the amplimers have hybridized to the target. These sites are the only double-strand regions in our reaction test tube. All of the rest of the DNA, except for our target is not copied! The synthesis extends along the DNA strand heading from 5′ to 3′ at about 1,000 bp per second. This extension is usually done at a slightly higher temperature, about 72°C. After a short while we have two double-stranded copies of our target DNA. We heat the test tube back up to 95°C, and the DNA again melts. Now we have four single strands of our target to which the amplimers can bind. We cool back down to 60°, rehybridize the amplimers, and then warm to 72° to extend. Bang! In a few minutes we have eight strands of target DNA. Every time we cycle between 60°, 72°, and 95°C, we double the number of copies of our target DNA. PCR

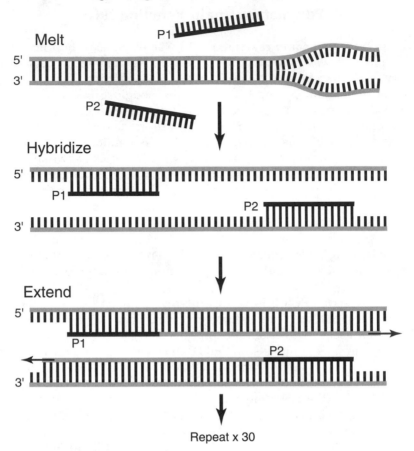

Figure 3.3. The polymerase chain reaction (PCR) consists of three phases—melting, hybridization, and extension—controlled by thermal cycling of the reaction mixture. Two synthetic amplimers, P1 and P2, bind on either side of our target region of interest. Each cycle doubles the number of DNA copies of the amplified region.

amplifies only our target. We started with a sample of DNA that contains the entire genome; after 4 hours we have an amplification product that consists of just one specific short piece. The target could be a portion of a gene, a part of a virus, or anything else that we wish to detect and amplify.

A PCR reaction usually involves 30 cycles and takes up to 4 hours. Figure 3.4 demonstrates the products of a PCR run separated on a gel by electrophoresis. The appearance of a target band of specific size means that the target DNA segment that we were looking for was present in the sample. PCR is carried out in a small box called a thermal cycler. A really neat feature that greatly simplified PCR was finding a DNA polymerase that did not mind being exposed to a temperature

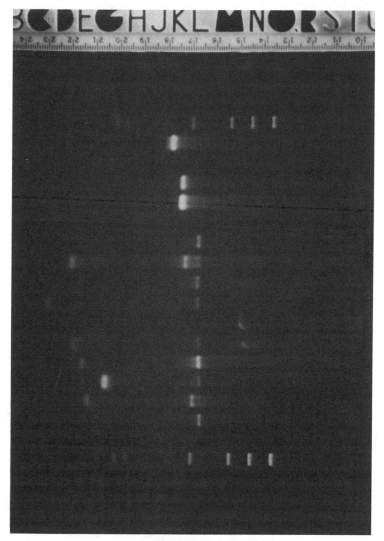

Figure 3.4. An agarose gel electrophoresis of the products of a PCR. Lanes 1 and 18 are molecular weight markers; lanes 2 to 17 show bands that represent amplification of DNA fragments. The bright bands toward the middle of most lanes represent amplification of a control gene. The fainter bands near the right side of the gel in some lanes indicate amplification of the target gene under investigation. (Courtesy of Dr. Steve Schichman, University of Arkansas, Little Rock.)

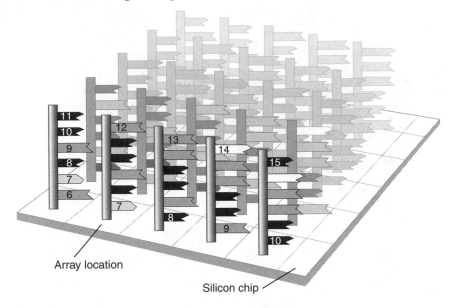

Figure 3.5. A schematic diagram of a small corner of a DNA microarray, also called DNA on a chip. Tens of thousands of short pieces of DNA are affixed at specific sites on the array.

near boiling! Almost all enzymes work best at 37°C and are denatured into an inactive form by boiling. That's why cooking works. Very smart people realized that the algae living in the hot geysers in Yellowstone National Park must have a DNA polymerase happy at near boiling temperatures. The enzyme cloned from these algae is called Taq (for *Thermal aquaticus*, the name of the algae). In Chapter 5 we will use PCR analysis to find patients with a genetic risk of thrombosis due to a mutation called factor V Leiden. Jump ahead if you wish to Fig. 5.2 to see a clinical example of PCR.

DNA on a Chip

The development of DNA on a chip merges molecular and computer technologies to handle large amounts of DNA sequence information. A DNA chip or **microarray** consists of thousands of synthesized oligonucleotide probes bound at specific spots on a chip substrate. The manufacture of these chips uses combinatorial chemistry to synthesize the probes one base at a time. The localization of the probes to a specific spot utilizes the techniques of computer chip fabrication with microphotolithography. DNA chips can have up to 1 million probes per square centimeter of chip surface. Figure 3.5 is a schematic of what one corner of a DNA chip looks like. A short piece of DNA is fixed to each spot in the array.

GeneChip, a product of Affymetrix (Santa Clara, CA), offers several arrays with about 30,000 probes on a 1.28 cm^2 chip. GeneChips have been designed for partial genotyping of HIV to detect mutations affecting drug sensitivity, mutational analysis of human tumor suppressor genes *p53* and *BRCA1,* and SNP analysis for drug discovery.

DNA on a chip analysis is similar to our other techniques, with the exceptional quality of probing for a large number of hybridizations all at once. Sample DNA is amplified to increase the amount of target, i.e., increase the amount of HIV or *p53* genetic sequences in a patient sample. The amplified DNA is labeled with a colored, chemiluminescent, or fluorescent reporter molecule. The DNA in solution is applied to the DNA chip under conditions to optimize hybridization. The chip is then examined to see where hybridization has occurred. The examination of the chip is by an automated microscope system that can look at the tens of thousands of probe sites.

Other companies and researchers have developed variations on this theme to compete with Affymetrix. These involve other means of creating the chip such as microjet application of presynthesized probes. Some chips use an electrical impedance system to detect hybridization. Many very clever automation and miniaturization methods are being combined to create an entire DNA laboratory as a microdevice (see McGlennen, 2001). Genomic information is now large scale; we frequently want to know the whole sequence of a gene rather than just detect a single mutation. As we look ahead to applications of these DNA microdevices, let's benefit from the experience of the computer revolution. We need not worry about the time and complexity of gene sequencing or other analysis. The technology is growing to handle the vast amounts of data that we seek to analyze.

Expression Arrays

To understand further the possibilities of DNA chip technology, let's consider an interesting and somewhat special application, **expression arrays**. As we discussed in Chapter 1, now that we have mapped the human genome, we are ready to move on to functional genomics. Expression arrays use DNA chip technology to map what a cell is doing right now. Cells are treated to extract and preserve all of the mRNA present. Recall that mRNA is very labile. It is a message, and, as such, is designed to degrade rapidly once it has been read. Numerous enzymes rapidly destroy mRNA, especially if it is released outside the cell.

When a tissue is harvested, a very large number of different mRNA molecules will be present. Each represents the transcription from a gene that is currently active. To return to our library analogy, looking at mRNA

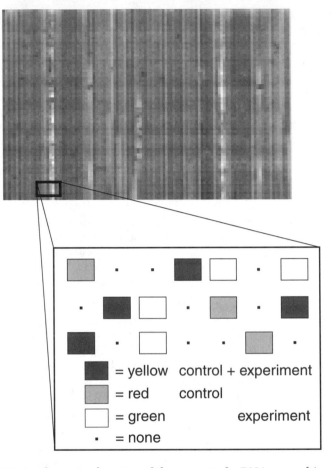

Figure 3.6. A schematic drawing of the output of a DNA-on-a-chip expression microarray. Each point shows the comparative gene expression of a tumor before and after irradiation (see text).

from a tissue is like looking at what books have been checked out from the library. For the moment we don't care what is in the library; we want to know what people are reading. Expression arrays tell us what a tissue is doing, by showing us all of the genes currently being transcribed. An expression array is a DNA chip designed to contain thousands of probes for genes that might be active in a tissue. We might choose these thousands of genes for our expression array based on the tissue we are analyzing. For the moment, even DNA chip technology cannot probe for all 35,000 human genes at once. Figure 3.6 is a schematic drawing of the output of an expression array. In this exercise, mRNA has been isolated from a control sample and from an experimental sample, for instance, a tumor before and after exposure to irradiation. Control mRNA is labeled with a red reporter molecule and experimental mRNA with green. At each of the 10,000 gene probe sites on the expression array, there is (0) no mRNA binding, (1) red binding, (2) green binding, or (3) both red and green binding (which looks yellow). An optical scanner and computer read the chip producing a color print out of the binding results. In Fig. 3.6 the colors have been reduced to four levels of white to black. The change in the expression of thousands of genes in the tumor sample can be analyzed. All the points in the array that are not black or white show a difference between gene expression in the tumor after radiation (see Freeman et al 2000; Yudung and Friend, 2001). This is a great way to see how tumors behave. An expression array signature such as in Fig. 3.6 might be a far more specific diagnosis than conventional histology of tumors. When we come to understand the data better, knowing the mRNA pattern of a tumor might tell us more about its aggressiveness as well as its sensitivity to chemotherapy.

Antisense Oligonucleotides

Antisense oligonucleotides are a new class of drugs that can modulate gene expression. They are short pieces of DNA, typically 15 to 30 bp, produced synthetically. The base sequence of an antisense oligonucleotide is made complementary to the sequence of a target mRNA molecule. An antisense copy is complementary to the mRNA that contains the genetic message in the correct "sense" orientation. Our choice of the length of the antisense and the exact portion of the mRNA we select as a target is critical. Too short (say 10 bp) an antisense could bind at other sequences besides our target. Under appropriate hybridization conditions the antisense oligonucleotide binds to its complementary portion of the mRNA and nowhere else. Figure 3.7 shows a short antisense molecule binding to mRNA. The bound antisense produces a short double-stranded sequence along the otherwise linear single-stranded mRNA. This short double-stranded region prevents translation of the mRNA into protein. This is because double-stranded RNA is recognized by the cell as abnormal and is destroyed by ribonuclease H. Also, if the antisense oligonucleotide is targeted to a region near the 5′ cap on the mRNA molecule, it will not be able to bind to a ribosome.

The result of antisense binding to mRNA is the functional destruction of that message. If an antisense oligonucleotide is present in excess concentration within the cytoplasm of a cell, no message can be successfully processed into protein. Thus, despite the fact that a specific gene is turned on and actively being transcribed into mRNA, the message will not get through. Antisense oligonucleotides are a very specific mechanism for interfering with the expression of only a single gene.

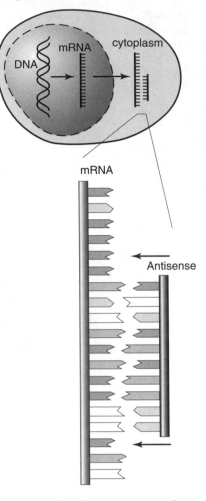

Figure 3.7. Antisense DNA binds to a portion of an mRNA molecule where it finds its match with complementary base pair sequences. The binding creates a double-stranded region on the mRNA that interferes with translation into protein.

The blocking of gene expression by antisense sequences can be made permanent, rather than dependent on repeatedly dosing the cell with more drug. This is accomplished by transfection of an antisense sequence into a cell using a **vector** (as we will discuss shortly). Transfection with antisense sequences is a means of genetic engineering that can produce transgenic plants and animals with resistance to specific pathogens. In Chapter 8, we will consider some specific examples of antisense therapy of human diseases.

Knockout Mice

A common experimental tool in molecular medicine is the **knockout mouse**. To study how a specific gene functions in an organism, we selectively inactivate that gene and then see what happens. This is usually done by inserting a piece of DNA that blocks the gene from functioning, creating a "null" allele. The DNA insert that blocks functioning is injected into a mouse embryonic stem (ES) cell. Since only a few ES cells will pick up the insert, we must have a selection method for finding these cells. This recombinant ES cell is then injected into a mouse blastocyst, which is in turn placed into a receptive female mouse. When successful, this method produces a strain of mice that breed true for the null phenotype of the knockout gene. The process is relatively straightforward. If you cannot do it in your own lab, many commercial services producing knockout mice are available. In Chapter 5, we will consider an experimental model of atherosclerosis in which we employ a strain of mice lacking the low-density lipoprotein (LDL) cholesterol receptor. If you want to know more, visit Boston University's Transgenic Core Facility Web site at *www.bu.edu /transgenic/knockout.html.*

Gene Vectors

A very necessary tool in DNA technology is a means of carrying a gene or smaller piece of DNA from one place to another. A gene vector is an engineered microorganism; virus, plasmid, bacteria, or yeast that can carry a foreign piece of DNA. Figure 3.8 demonstrates how a vector works. The vector here is a pBR322 plasmid, an obligate parasite of *Escherichia coli* bacteria. The native DNA of pBR322 is a circular loop that can be cut with the restriction enzyme Hind III. The cut leaves sticky ends. We make use of these sticking ends to ligate our gene fragment into the loop, resealing it. To do this, we attach leader and trailers to our gene to match the ends of the Hind III cut. In Fig. 3.8, our insert is a fragment of the myeloperoxidase (*MPO*) gene. Incubating cut plasmids with our insert DNA results in a few plasmids resealing with our insert incorporated into their genome. Many plasmids reseal without the insert; we need a selection method to get rid of empty plasmids. A typical selection method is to have our insert also carry a gene for antibiotic resistance. We mix our plasmid with a broth of *E. coli* bacteria, and then plate the broth out on a Petri plate containing the antibiotic. The only colonies that will grow are derived from *E. coli* that have picked up a plasmid with an antibiotic resistance gene coupled with our insert gene. We now have a bacterium that we can grow to produce a large amount of our gene. We can give samples to other labs so that they can have a copy of our gene. Genetic engineering uses many dif-

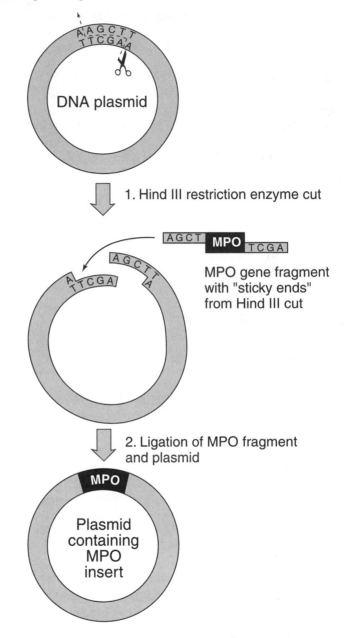

1. Hind III restriction enzyme cut

MPO gene fragment
with "sticky ends"
from Hind III cut

2. Ligation of MPO fragment
and plasmid

Figure 3.8 An example of gene cloning demonstrates the step of inserting a piece of the *MPO* gene into a plasmid.

ferent kinds of vectors. Some, like the yeast artificial chromosome (YAC), are designed to carry very large segments of human DNA and are especially useful to the Human Genome Project. Others, like adenovirus, are designed to be infectious to human cells, and are called cloning vectors, capable of inserting a gene into a human tissue. Table 3.1 lists some of the common cloning vectors.

Cloning Vectors and Gene Therapy of Cystic Fibrosis

As a nuts and bolts example of the use of vectors, let's consider the challenge of gene therapy in the treatment of cystic fibrosis. We wish to insert a copy of the *CFTR* gene (to be discussed in more detail in Chapter 5) into the respiratory epithelium of patients with cystic fibrosis. If successful, this would correct the defect in the membrane transport that causes the respiratory problems of this disease. To begin, we will use the AdV adenovirus vector. AdV is a DNA virus that produces infections of the upper respiratory and gastrointestinal tracts. These are common infections producing illness in millions of people yearly. To use AdV as a vector, we must modify it to make it incapable of viral replication, and thus prevent it causing an illness in our patient. Initially, the most common step was to remove the E1 region of the viral genome. This vector, called ΔE1 AdV, is easy to grow in culture. The virus can accommodate a large insert up to 30 kilobase (kb) of a gene package. This vector has been used in clinical trials of gene therapy of cystic fibrosis. The *CFTR* gene with a promoter in front is inserted into the vector, which is then applied to the respiratory epithelium of the patient. The AdV viral package is taken into the cell by binding to a receptor on the epithelial cell that subsequently undergoes endocytosis. There it is, our corrected gene is in the target cell!

The problems with AdV and similar vectors are (1) host immune response that fights the vector, producing toxicity and limiting repeated doses; and (2) limited duration of the inserted gene expression. Obviously, we cannot give bronchitis or pneumonia to a cystic fibrosis patient. Engineering steps to decrease the host response to AdV begin with deletion of the E4 region of the virus. This limits the number of viral proteins expressed by the vector and minimizes the immune response. This works. In fact researchers found that they could remove essentially all of the viral genes from AdV, creating a "gutless" AdV vector with no potential for viral replication and limited immune response from the host. This still leaves problem number 2; the inserted gene is not expressed for very long. As the respiratory epithelium renews itself every few weeks, all of the transfected cells are sloughed. To continue the effect, we must re-dose the patient. Cystic fibrosis is a chronic disease; weekly very expensive and complex therapy is not a great option.

Table 3.1. Examples of cloning gene vectors.

AdV, adenovirus—easy to produce; takes large insert; "gutless" variant invokes less immune response; does not integrate into genome, resulting in limited duration of transgene expression

AAV, adeno-associated virus—cannot replicate without co-infection by HSV or AdV; integrates into genome; can only accommodate small inserts (4.5 kb)

HSV, herpes simplex virus—especially good for reaching targets in the central nervous system; does not integrate into genome

Liposomes—lipid particles containing DNA; easy to produce in large quantities; do not invoke immune response; do not easily integrate into genome

To solve the problem of limited duration of transgene expression, we move on to vectors that integrate into the genome. If an integrating vector could infect the basal cells of the epithelium that generate the surface respiratory epithelium, we might get a very long term effect. Adeno-associated virus (AAV) is a good candidate for a vector; it naturally inserts at a specific site on chromosome 19. AAV, after inserting on chromosome 19, normally waits for a co-infection with a second virus such as AdV or herpes simplex virus (HSV) before it replicates. AAV by itself does not replicate. Again for safety and to reduce the immune response, we engineer a gutless AAV vector. One limitation of AAV is that it is a small virus and can accommodate only a 4.5-kb insert. This is a problem for large genes such as *CFTR*.

Other vectors including nonviral packages like liposomes all have potential uses, but each carries its own engineering problems. The death of a patient receiving an experimental gene therapy at the University of Pennsylvania in 2000 was due to unexpected toxicity of the vector, producing hepatitis.

To date (West and Rodman, 2001) more than 20 clinical trials of gene therapy for cystic fibrosis have been conducted using all of the vectors in Table 3.1. Lung inflammation as a toxic side effect has been significant in some studies. No effect or very limited effect of the inserted gene has been found. The early promise of gene therapy has been dulled by the experience of these and other trials. Yet these are engineering problems, and genetic engineering as a discipline is in its infancy. I am confidant that engineering problems can be overcome and gene therapy will become a very powerful tool of medicine.

Cloning

Ten years ago when writing the first edition of this book, **cloning** implied copying a gene. That was the cutting-edge technology back then. A decade later, copying a gene is a high school science fair project.

When we say cloning now, we mean making multiple copies of an entire complex organism; a sheep, cow or possibly a human.

Cloning a Gene

Since cloning a single gene has become so routine, let me give just a brief example. I wish to clone the gene for myeloperoxidase (*MPO*). This is an enzyme present in neutrophils that aids in killing bacteria. Let's say that I wish to clone *MPO* so that I can modify the gene for use as an antibacterial agent in a special application. To begin, I have to find the gene sequence for *MPO*. In past years, that would generally involve finding the gene. Now, the human genome is sequenced, so I begin by consulting a gene database. I find the entire *MPO* gene sequence (LocusLink 4353, OMIM 254600) is already cataloged.

To get a copy of *MPO*, I choose to probe a cDNA library. (An alternate approach might be to ask for a copy from someone that already has an *MPO* clone. For the purpose of this example, let's pretend that we want to do it ourselves.) I begin by synthesizing a short DNA oligonucleotide complementary to a portion of the *MPO* gene based on the published sequence. To save time and money, since my lab does not have a DNA synthesizer, I will order my *MPO* probe on the Internet. A 20-bp piece of DNA made to my exact specifications is inexpensive and available in a day or so. Next, I acquire a cDNA library for blood cells, again obtained from a friend or a commercial source. This too is not expensive. A cDNA library is a collection of unsorted DNA pieces. These cDNA pieces are made by isolating all of the mRNA produced by a sample of blood cells, then copying these mRNA back to DNA using an enzyme called reverse transcriptase.

The fragments of the cDNA library I have chosen are packaged within plasmids, one of the gene vectors that we discussed. I will use my small oligonucleotide probe to find which of the plasmids in my library contain the *MPO* gene sequences. I do this by hybridizing my probe to a Petri dish plated with *E. coli* containing the many different plasmid inserts of the library. My probe sticks to only one or two colonies. I lift these colonies and grow them in broth. I now have plasmids containing only *MPO* sequences.

From this point, what I do depends on my needs. For this demonstration, let's say that I want to modify the *MPO* gene to produce a cloned variant that in the right situation expresses a large amount of the protein. I want to make a biologically active antibacterial agent that I can turn on or off. I would need to clip and rearrange my gene pieces to get the active part of the gene assembled. I will then add a strong promoter segment in front of my gene. The promoter will turn on my cloned *MPO* in response to an external signal, giving me control of gene expression. Finally I place my entire "construct," my super *MPO* gene, into an appropriate vector. I might choose the vector lambda gt11.

This plasmid is considered a good choice for expressing a gene product in a host cell. I now have a cloned super *MPO* gene in an expression vector. I can grow very large, even industrial, amounts of my gene in a vat of *E. coli* and broth.

I might at this point wish to consider whether anyone already "owns" this gene, especially if I am planning a commercial use for my clone. Who owns a gene is a new legal, moral, and commercial question. My original MPO sequence may have come from a public library without restriction. Alternatively, I might have acquired the information from a corporate library that will require a licensing agreement when I develop a product.

Cloning a Human

Cloning a human can and will be done. This is true even though I believe human cloning should not be done. Figure 3.9 schematizes the steps involved and will serve as an outline for our discussion.

Choice of Donor Cell

The human to be cloned donates a cell. We will use an 'adult' cell. An adult cell is any cell in the body other than an **embryonic stem (ES) cell** or a **germ cell**. The germ cells, we already know; they are the haploid sperm and oocyte. Sperm and oocyte cannot be donor cells for cloning because they contain only one copy of each chromosome and only one sex chromosome, X or Y. Embryonic stem cells are a few cells set aside during early embryo development for generation of germ cells to procreate the next generation. When a human fetus is only days old, it already has established a reserve of embryonic stem cells. Embryonic stem cells are an ideal source for cloning: they are diploid, and they are primed for development. ES cells were the donor cells in early animal cloning experiments.

However, acquiring ES cells is difficult. A major controversy exists over whether research using human ES cells should be allowed. ES cells are useful for many things besides cloning. They are a potential source of new neurons for stroke patients and patients with spinal cord injuries. ES cells can likely generate new heart muscle cells for patients in heart failure. The sources of human ES cells are (1) aborted human fetuses, (2) unutilized embryos from in vitro fertilization donors, and (3) placental or other tissues. You can see the ethical problem. Current thinking is that sources (2) and (3) may be OK but (1) is not.

This brings us to the use of adult cells for cloning. An adult cell is any other cell besides an ES cell and a germ cell. We use adult cells because ES cells are hard to come by, especially from an adult human.

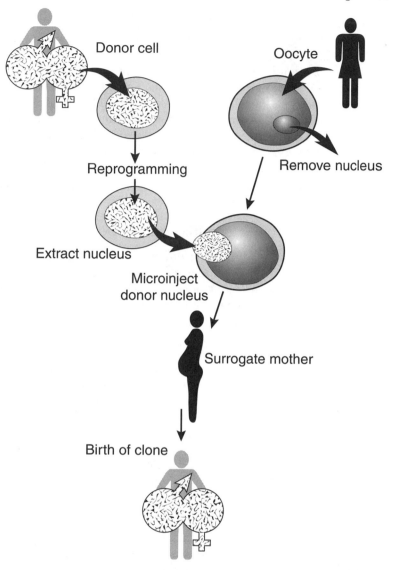

Figure 3.9. Human cloning involves a series of steps that include inserting DNA from the selected adult cell into an ovum that has had its own DNA removed. The ovum is then implanted into a surrogate mother.

In Vitro Fertilization (IVF)

There is an important distinction between cloning and **in vitro fertilization (IVF)**. IVF does not create a genetic copy or clone. IVF combines the two haploid germ cells of a male and a female outside the body. There is no new genetics here, just a mechanical assistance to the process of fertilization. The zygote is implanted in a surrogate womb. The surrogate mother usually is also one of the biological parents. The germ cells in IVF are the product of meiosis and are programmed for combining as a zygote and for the fetal development that follows. Occasionally the fertilized zygote splits into two or more cells. This occurs in natural reproduction as well as IVF and is the source of identical twins, who are, in fact, clones.

Reprogramming

The adult cell that we will take from our donor must be reprogrammed before it can begin division and growth as a fetus. This step is shown as a box in Fig. 3.9. It represents a complex treatment of the adult donor cells. Reprogramming means undoing cell changes such as chromosome condensation, masking of genes, and other things that we do not understand. This step was until recently a barrier to cloning from adult cells. Reprogramming is also the source of some very serious problems with cloning. In animal experiments, the viability of embryos produced by cloning of reprogrammed adult cells is only a few percent. This means that many cells have to be processed and most discarded. The animal clones that survive until birth have some unexpected problems. They are frequently abnormally large fetuses that potentially threaten the health of the surrogate mother. Obesity develops in young cloned animals. Developmental problems of the immune and central nervous systems also occur. At the moment, scientists think that these problems come from the reprogramming of adult cells.

Genetic engineering of the clone

The first steps of embryogenesis of our donor cell are a point for further intervention if we want our clone to be something different than an identical copy. This would be the step to add any transgenic material, such as correcting a mutated cystic fibrosis gene (see Chapter 5). We could also introduce transgenic vaccines at this stage. This would make the clone resistant to diseases such as HIV and human papilloma virus (HPV) (see Chapter 4). At the very least we would divide the early embryo into a group of cells, to have multiple clones in reserve.

Fertilization

The nucleus of the donor cell is injected into an oocyte that has been stripped of its own nucleus. The donor cell supplies all the DNA and the oocyte supplies the environment. Human oocytes are not hard to come by; they are available as extra eggs harvested from IVF procedures.

Implantation and Birth

As the embryo develops, it becomes ready for implantation into a surrogate mother. We do not have the capability as yet to support the growth of a mammalian fetus outside the womb. The surrogate mother provides the environment for fetal development.

The fetuses born of human cloning will be approximately the equivalent of a genetic twin of the donor and of each other (if multiple clones are produced). The clone will differ in having mitochondrial genes inherited from the oocyte, not the donor. Identical twins share the same womb; this will not be the case for our clone. In Chapter 1, I mentioned that there is much that the genome does not code for, particularly the brain, which is the result of self-organization and development. Thus, clones will develop differently, beginning even before birth.

The reason for cloning an animal is to engineer the clone for research studies or to produce a variant that is in some way better. The reasons put forward for cloning a human are quite varied (see Talbot, 2001). The reasons include recovery of a dead child, a second chance for an adult, and manufacture of spare parts for transplantation. The extent to which making a clone that is close to an identical twin of the donor satisfies these reasons is very uncertain.

Aside: Human Spare Parts

Cloning a human is an extreme test of ethical, philosophical, medical, and technical issues. The ethical question alone is overwhelming. The problem will be met in many forms, beginning in partial human cloning. We already have inserted human genes into laboratory animals and plants. We have no apparent problem with a tobacco plant producing a human protein that breaks up blood clots. Nor does a lab mouse with human hemoglobin S, a model of sickle cell disease, seem to be an issue. As we create animals with more and more human traits, our sense of concern may increase.

Let's jump ahead to the issue of a full human clone done for a medical reason: replacement tissues and organs. What if we determined that the best treatment for a 10-year-old with cystic fibrosis (CF) was creating a cloned twin with a reengineered *CF* gene that was not defective? Then as the child aged we would have a set of good replacement lungs and pancreas from a 10-year-younger identical twin. Is it morally unacceptable and in fact murder to remove vital organs from the younger child to save an older sibling? What if the clone is never truly born?

The clone could be produced as an anencephalic (brain absent) fetus implanted in a nonhuman surrogate mother and raised in a tissue **pharm**. The spare part clone would never be a conscious entity. To me this is repugnant and unacceptable. It is, however, on the verge of being feasible.

In Chapter 8, I will return to this issue of replacement tissues. ES cells obtained from discarded placentas may be a means of curing diseases by regeneration. Laws banning cloning may be written too broadly, also prohibiting the use of any type of stem cell. Quite literally, we need to pay attention to both the baby and the bathwater in making these regulatory decisions.

Summary

The molecular tools available to us allow almost any manipulation of the genome. We can sequence even large pieces of DNA very rapidly. As we learn the function of genes, we will also be able to manipulate their function through antisense or direct gene modification. Cloning is the process of copying and modifying genes. We can even clone an entire organism, making multiple copies of its genome. This chapter completes the study of the tools of molecular medicine; we will now look at applications in many different areas of medicine.

Bibliography

Freeman WM, Robertson DJ, Vrana KE. Fundamentals of DNA hybridization arrays for gene expression analysis. *BioTechniques* 2000;29:1042–1055.

McGlennen RC. Miniaturization technologies for molecular diagnostics. *Clin Chem* 2001;47:393–402.

Mulllis KB. The unusual origin of the polymerase chain reaction. *Sci Am* April 1990, pp. 56–65.

Talbot M. A desire to duplicate. *New York Times Magazine* Feb 4, 2001, pp. 40–45.

West J, Rodman DM. Gene therapy for pulmonary diseases. *Chest* 2001;119:613–617.

Yan H, Kinzler KW, Vogelstein B. Genetic testing—present and future. *Science* 2000;289:1890–1892.

Yudong DHe, Friend SH. Microarrays—the 21st century divining rod? *Nature Med* 2001;7:658–659.

PART II

Molecular Approach to Disease

4   Infectious Diseases

Overview

In Part I we considered the basic principles and technology that underlie molecular medicine. This chapter on infectious diseases begins Part II. We will apply the basic principles that we have learned to the study, diagnosis, and treatment of disease, using specific clinical examples. For infectious diseases, the application of molecular medicine is both straightforward and dramatic. Reflect for a moment on our current methods for the diagnosis and treatment of an infectious disease. A patient presents with signs and symptoms that, although nonspecific in nature, suggest an infectious process. These initial signs we know to be due in part to the body's inflammatory response and in part to the toxic effects of the infectious agent itself. The touchstone of diagnosis is the identification of the bug. This is mostly accomplished by culturing it in the laboratory. Broths, Petri plates, and biochemical tests are the standard tools of the microbiology lab. Now consider the possibilities of molecular medicine. A clinical sample of sputum, blood, or urine is analyzed for nonhuman DNA. DNA (or RNA) allows us to identify a microorganism without culture. Molecular medicine will also dramatically alter our therapy of infections. We are pushing antibiotics to their limit, fighting acquired multidrug resistance. Molecular therapies include antisense, ribozymes, DNA vaccines, and immunotherapy. The distinction between drugs against the bug and drugs that alter the immune system may become slight as we combine both types of therapy.

Microbial Genomes

The human genome is 1,000 times larger than a typical bacterial genome and 10,000 to 100,000 times larger than a viral genome. However the number of clinically relevant bacteria, viruses, parasites, etc.

is very large. Cataloging the DNA sequences of important microbes is therefore a very great task. By the end of the year 2001, hundreds of organisms have been fully sequenced, beginning with *Haemophilus influenzae* and now including *Helicobacter pylori, Treponema pallidum, Escherichia coli, Mycobacterium tuberculosis,* and other favorite human pathogens. Many more organisms will be genetically sequenced in the next few years.

What have we learned from this rapidly growing body of data that is of use in treating human infectious diseases? First of all, once we have part of the genetic sequence of an organism, we can identify it with DNA probes. This will, I believe, replace standard culture methods. Knowledge of an organism's genome also gives us the potential to produce specific antisense drugs to block the growth of that organism. Alternatively, we may use our knowledge of the gene sequence for a specific protein to create a DNA vaccine against that organism. Microbial genomics also teaches us how organisms achieve virulence—that is, their ability to overcome rather strong host defenses. To know an enemy is the first step in defense. This chapter discusses each of these topics individually. We will see how knowing the genome of an organism gives us many new tools to limit its pathogenicity.

Molecular Detection of Microorganisms

Every microorganism has genomic sequences that make it unique. To develop clinically useful molecular probes for infectious diseases, it is necessary to decide which species of organisms you wish to identify. A probe can be made very specific for a single strain of one bacterial species, or a probe can have much broader specificity, encompassing many species. For example, it would be possible to make a probe to hybridize to DNA sequences common to all coliform bacteria. It would also be possible to become much more specific and find DNA sequences that are unique to enterotoxic *E. coli.* For screening, less specificity is desirable. A test to screen for infection should include a wide range of possible organisms, perhaps somewhat focused by the symptoms and physical findings. We could construct a DNA chip to probe for a wide range of infection associated diarrheal diseases, or alternatively for a range of upper respiratory infections. The chip or other molecular technique would require a mix of many probes to bacteria and viruses. A positive result from a screening method might then key off a much more focused application, such as typing the strain of a *Salmonella* species implicated in an outbreak of food poisoning.

Infectious disease probes can be directed against the DNA or RNA genome of the microorganism; however, in some instances, other components of the microbe serve as better targets. For example in bacteria, ribosomal RNA, which is present as a large number of copies per bacterium, can be a better target than DNA, which is present in only a single copy. Our molecular identification of microorganisms can also be extended to characterize their pattern of antibiotic resistance. Antibiotic resistance is part of the genome of bacteria, frequently carried within plasmid DNA sequences.

The sensitivity of molecular techniques for the detection of microorganisms is usually very much greater than standard cultures. A polymerase chain reaction (PCR) method will give a positive result if only a few microorganisms are present. Does a single mycobacterium in the sputum indicate tuberculosis? Clinical experience with adjusting the sensitivity of molecular techniques to what is significant, with additional experience in the interpretation of the data, is required.

Molecular detection methods can use many forms of hybridization depending on the specific application. In-situ hybridization allows a color labeled probe to be visualized in a microscopic section. Look ahead to Fig. 7.8, which demonstrates an in-situ hybridization for human papilloma virus (HPV). In this application we want to see what tissues demonstrate persistent presence of virus, as this poses a risk for the development of cervical cancer. Other hybridization methods for detection of infectious agents include chemiluminescence or ferromagnetic beads in liquid suspension. Precisely localized oligonucleotide pieces on a solid substrate (DNA chip) offer the greatest range of multiple probes. DNA probes for microorganisms can not only identify an organism but also recognize specific strains that will help trace the source of exposure. DNA probes can also identify drug-resistant forms. Future automation of the molecular detection of infectious diseases should very significantly reduce the cost of testing while also greatly reducing the time for identifying microbes. As with many other areas of molecular technology that we will consider in this book, the greatest limitation will be the pace of our ability to adapt clinical experience and practice to these new tools.

Miniexample: Mycobacteria

Molecular detection methods in hospital microbiology laboratories are currently mostly applied to the detection of organisms that are difficult or slow to grow, such as mycobacteria. *M. tuberculosis* organisms can be detected in several hours by molecular probes, whereas standard culture can take 3 to 6 weeks to produce a result. Two Food and Drug Administration (FDA)-approved nucleic acid amplification methods for TB have proven useful: Gen-Probe (San Diego, CA) and Amplicor (Roche, Branchburg, NJ). DNA microarrays for mycobacteria detection are in development. Microarrays have the potential to detect the bacteria, identify the species, and detect mutations associated with rifampin antibiotic resistance, all in one rapid assay.

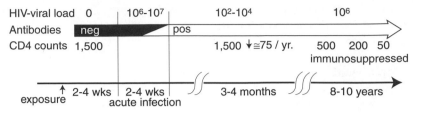

Figure 4.1. The time course of HIV infection is characterized by an acute phase with rapid proliferation of virus. The immune system produces antibodies by 6 weeks and the viral load drops. A long chronic phase follows, with the virus held in check by the immune system. Eventually a loss of CD4 lymphocytes (about 75 μl/year) leads to an immunosuppressed state.

HIV: Molecular Monitoring of an Infection

HIV is the prototype of a molecular infectious disease. The mysterious symptoms and inexplicable epidemiology initially designated as the acquired immune deficiency syndrome (AIDS) is now known to be a disease caused by a virus. HIV can be detected and quantitated, and its spread tracked by molecular diagnostic tests. HIV and molecular medicine have grown up together. Without molecular technology, we would not have discovered the virus. Without the urgency of the worldwide HIV epidemic, molecular medicine would not have grown so fast. The standard tests for monitoring HIV infection are the most advanced molecular techniques in the clinical laboratory.

The time course of untreated HIV infection, although quite variable, is summarized in Fig. 4.1. Two to four weeks after exposure to the virus, an acute mildly symptomatic infection is manifest. At this point the virus has rapidly multiplied with little inhibition from the immune system. Very high numbers of HIV RNA viral particles can be measured in the blood, as high as 10^6 to 10^7 /μL. Within a few weeks the infection subsides, antibodies appear in the serum, and the viral load falls rapidly. By 3 to 4 months after infection, the viral load will be at a nadir, typically from 10^2 to 10^4 /μL. The CD4 helper T lymphocyte is the target cell of the HIV virus within the immune system. The number of CD4 cells is normal for a long time after infection. The battle between the virus and the immune system might lower the CD4 count by 75/μL each year of infection. Eight to ten years later, the CD4 count falls to a level that produces severe immunodeficiency. The opportunistic infections and cancers of AIDS occur at this stage.

Molecular testing is essential in the monitoring of treatment of HIV infection. The tests currently available are listed in Table 4.1. A key test is measurement of viral load. The initial viral load is predictive of the length of the chronic asymptomatic phase of AIDS. Changes in viral load also dictate the initiation and alteration of drug treatment. Beyond

Table 4.1. Molecular testing useful in HIV infection.

HIV serology—detection by enzyme-linked immunosorbent assay (ELISA), confirmation
             by Western blot
CD4 helper T-lymphocyte count—monitors degree of immunosuppression
HIV RNA viral load—predictive of time course of infection, monitors response
             to or failure of drug therapy
HIV genotypic assays—determines drug sensitivity, HIV strain typing for epidemiology
HIV phenotypic assays—when necessary, confirms drug sensitivity

counting the number of viral particles, we can sequence its 9.2-kilobase
(kb) genome. HIV has no proofreading error-correction mechanism
when it undergoes RNA to DNA copying. This leads to a high rate of
mutation that results in the emergence of drug resistance. HIV genotypic
assays using DNA chip or other technology identify specific strains of
HIV and their likely pattern of drug sensitivity. HIV phenotypic assays
are more time-consuming but can provide direct proof of the drug sensitivity of a virus isolate from a patient. Virus is grown in the presence of
various drugs, and suppression of viral replication is quantitated.

The HIV pandemic has commanded our attention. More complete
discussions of every aspect of this disease are available elsewhere. My
brief presentation here is to demonstrate the molecular technology that
has been implemented in the diagnosis and treatment of HIV. Other
chronic viral infections will benefit from similar testing. The monitoring of hepatitis C (HCV) already uses similar tools of viral load and
HCV genotyping.

Molecular Therapy of Infectious Diseases

Molecular medicine gives us many new tools with which to fight infectious diseases—just in time, in my opinion. The emergence of new infectious agents with greater host range and increased antibiotic resistance is frightening. We will consider briefly some of these molecular
tools. These are all at various stages of research and early clinical trials.
Table 4.2 summarizes the characteristics of these new molecular tools
against infectious disease, each of which we shall look at now.

Antisense

Antisense oligonucleotides were introduced in Chapter 3 as a means of
interfering with gene expression. We can just as easily use them to interfere with viral genes, especially genes for viral replication. Recall
from Figure 3.7 that what we need is a short piece of our engineered
RNA in the cell cytoplasm. How we get it there is the current problem
with antisense. If we give antisense as an IV drug, enzymes in the
serum rapidly degrade the RNA. We can chemically stabilize our engi-

Table 4.2. Molecular tools for fighting infectious diseases.

Antibiotic—drug that interferes with microbial biochemistry, e.g., penicillin interferes
   with bacterial wall synthesis

Antisense—RNA molecule that blocks a microbial gene, e.g., experimental
   HIV therapies

Ribozyme—RNA molecule that catalyzes the destruction of a microbial gene, e.g.,
   research on HIV, HCV

Anti-infectious proteins—protein that blocks microbial metabolism, e.g., soluble
   CD4 receptor protein binds to HIV virus before it attaches to lymphocyte; proteins of
   innate immunity on mucosal surfaces—defensins

Intracellular single chain antibodies, intrabodies—intracellular antibody, e.g.,
   anti-erbB2 oncoprotein antibody

DNA vaccine, e.g., anti–human papilloma virus vaccine

---

neered RNA to overcome that problem. However, it is a large molecule
that does not easily cross the cell membrane. We can carry it across the
membrane, enclosed in a package that the cell will take up, such as a liposome. A better approach is to carry the antisense across the cell
membrane as part of an expression vector. We construct a vector that
has an engineered piece of DNA that, when transcribed, produces our
desired antisense RNA. The vector binds to receptors on our target cell,
for instance the CD4 receptor on helper T lymphocytes. These cells
take up the vector, and the antisense becomes expressed in their cytoplasm interfering with HIV replication. A different vector might be
used if our target were hepatocytes and HCV. This approach gives us a
stable intracellular expression of an inhibitory antisense RNA molecule. The distinction between an expression vector containing antisense sequences and a DNA vaccine (which we will consider in a moment) is slim, but important. In **DNA vaccines**, we are transferring the
gene sequence for a piece of viral antigen to sensitize the immune system. Antisense inhibition of virus does not use the immune system to
carry out viral killing. A further extension of the use of antisense is to
add the expression vector to the germline of a plant or animal that we
wish to make permanently resistant to a virus. This is one of the features now added to certain genetically engineered plants.

Ribozymes

An enzyme is a protein encoded by a gene that catalyzes a biochemical
reaction. A **ribozyme** is a piece of RNA that catalyzes a biochemical reaction. Ribozymes were discovered long after enzymes. The possibility
that RNA could be anything other than an intermediary between DNA
and protein was ignored. This was contrary to the central dogma of the
early days of molecular biology, which stated that information flowed

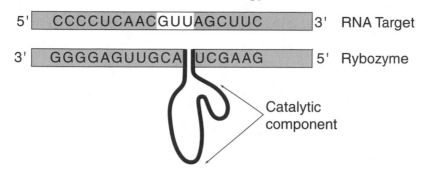

5' CCCCUCAACGUUAGCUUC   3' RNA Target

3' GGGGAGUUGCA UCGAAG   5' Rybozyme

Catalytic
component

Figure 4.2. Ribozymes are short pieces of RNA that bind and degrade targets
on RNA molecules.

from DNA to RNA to protein. We now have reasons to believe that in the
evolution of life RNA genes and ribozymes preceded DNA or protein.

We can use ribozymes as therapeutic agents against infectious diseases. RNA molecules are easier to engineer and produce than protein
molecules. Figure 4.2 shows an engineered ribozyme. A portion of the
RNA is constructed like an antisense, complementary to a target sequence on another RNA molecule. The binding between the two complementary sequences holds the ribozyme in place. At one particular
triplet in the middle of the target sequence, the catalytic component of
the ribozyme cleaves the target strand, thus permanently inactivating it.

Ribozymes (like enzymes) are not consumed. When they break one
strand of target, they are available to move off and find another, and another, etc. Ribozyme binding has very high specificity; we can be certain of our target with little "collateral" damage. Ribozymes are made
of RNA and when injected are therefore not antigenic (as protein molecules would be). Unfortunately, ribozymes, like all native RNA molecules, are rapidly degraded in the body. We must find a means of stabi-
lizing ribozymes to withstand enzymatic degradation in the serum, to
make these molecules useful therapy.

Antiinfectious Proteins

The administration of an immune "serum" is one of the oldest immune
therapies. The Alaskan Idiatrod sled dog race is a commemoration of
the 1925 race to transport diphtheria antiserum to Nome. We only occasionally use immune globulin therapy now; examples would be bot-
ulinum antitoxin or snake antivenom. However, with molecular technology, it would not be difficult to manufacture large quantities of pure
immunoglobulin specific to an antigen. Antiserums may once again be
a useful tool.

Intracellular Single-Chain Antibodies—Intrabodies

Manipulation of the antibody gene family permits engineering of a gene construct that encodes a modified antibody molecule. This antibody molecule consists of a single amino acid chain and is expressed within the cytoplasm of the cell. These artificial antibodies are called **intrabodies**. Their intracytoplasmic expression makes them effective against targets that would not normally be seen by immunoglobulin antibodies. Recall that normal immunoglobulins are bound to the surface of lymphocytes or free in the serum. Intrabodies can attack cytoplasmic proteins such as oncoproteins or HIV components.

DNA Vaccines

DNA vaccines stimulate the major histocompatability complex (MHC) I pathway of the immune system, the major effector cells of which are CD8 suppressor T lymphocytes. DNA vaccines heighten cell-mediated immunity. Conventional vaccines stimulate humoral immunity and the production of soluble antibodies (immunoglobulins) through the MHC II pathway. Conventional vaccines use as an antigen either live attenuated or killed microorganisms, or purified protein from the coat of the organism. DNA vaccines achieve MHC I response by presenting the antigen intracellularly. DNA vaccines are new because until we developed molecular technology, presenting the antigen intracellularly was not easy to do. DNA vaccines consist of a vector that carries a piece of the genome encoding our antigen. The vector carries this piece of DNA into the cell, and causes the gene fragment to be expressed. That means that the fragment needs its own promoters, or other means for turning itself on. Expression of the vaccine gene fragment means that the DNA is transcribed to RNA that is in turn translated into protein. This protein is the intracellular antigen that stimulates MHC I cell-mediated immune response. Another method for getting the engineered DNA fragment into the cell besides a viral vector is use of a "gene gun." DNA fragments can be shot into the skin with a certain percentage of DNA pieces ending up within the cells. Whatever the method, the key is to get a fragment of DNA into a cell in such a way that it will be transcribed into a protein fragment that will be recognized as antigen. Table 4.3 summarizes the differences between DNA vaccines and conventional vaccination. Table 4.3 is an oversimplification that assumes for purposes of illustration that a vaccine is either DNA or conventional. In fact, vaccination by DNA fragments shot into muscle by the gene gun produces both MHC I and MHC II responses of varying degrees of intensity. Conventional vaccines using attenuated live viruses also cross over and can produce some degree of cell-mediated immunity. The immune system is very complicated.

For many diseases, like measles, mumps, or influenza, conventional vaccines work. For other diseases, such as HIV, malaria, or prolonged

Table 4.3. Differences between DNA and conventional vaccines.

DNA vaccine	Conventional vaccine
MHC I cell-mediated immunity	MHC II associated humoral immunity
CD8+ suppressor T lymphocytes	CD4+ helper T lymphocytes and soluble antibody
Antigen is protein, intracellular	Antigen is protein, extracellular
Vaccine is delivered by vector or gene gun	Vaccine is injected; SC, IM, IV, or oral administration
HIV, HCV, HPV, malaria	Measles, mumps, flu

HCV, hepatitis C virus; HIV, human immunodeficiency virus; HPV, human papilloma virus; MHC, major histocompatability complex.

protection for influenza, we need a cell-mediated immune response as produced by DNA vaccines. The major need for DNA vaccines is our desire to stimulate cellular rather than humoral immunity. However, other advantages of DNA vaccines are important. They are stable at room temperature and do not require refrigeration. Conventional vaccines usually require refrigeration or are labile after reconstitution. DNA vaccine production uses a similar technology for each vaccine; we simply change the inserted piece of DNA that codes for the antigen. Conventional vaccines are each quite different in how they are prepared.

We will now consider a few examples of treating infectious disease with our new molecular tools. These therapies are all in the developmental or research stage, yet we can see their power and their future potential.

Example—Influenza Virus

The ubiquitous influenza virus, the cause of yearly epidemics of discomforting illness and sometimes mortality, is a negative segmented single-strand RNA virus. The virus rapidly mutates, changing its highly polymorphic envelope proteins. Vaccines that generate neutralizing antibodies against these proteins are effective prevention. However, the vaccines must be produced yearly, anticipating the strains most likely to cause disease that season. Revaccination with an updated vaccine is necessary every autumn. In addition to vaccination, antiviral drugs such as amantadine and rimantadine or neuraminidase inhibitors such as zanamivir if administered soon after symptoms or exposure may limit the illness. When flu develops, patients turn to over-the-counter symptomatic relief, visits to doctors, and requests for antibiotics.

All of this is a big-time industry to combat an illness that disables millions and occasionally kills. Let us consider, as an example of new molecular therapies, the possible production of a DNA vaccine against influenza. The goal of a DNA vaccine would be to induce long-term

protection against a wide variety of strains of influenza, making vaccination a one-time affair. For a DNA vaccine against influenza, our
choice of antigen would be some part of the influenza viral genome
that is highly conserved between viral strains. We will stay away from
viral coat protein genes, as this is the highly mutated portion of the
genome that confounds conventional vaccination. Good candidates for
conserved genes of influenza virus are hemagglutinin, neuraminidase,
and nucleoprotein-related genes. Several studies have shown that DNA
vaccines made with plasmids containing one of these genes confers
protection against a large viral challenge in mice, cattle, chickens, and
hogs. As an aside, I believe that physicians should be aware that molecular diagnosis and therapy of infectious diseases is very much more
advanced in veterinary medicine than human medicine.

With additional experience and clinical trials, I believe that DNA
vaccines for influenza will at least augment if not replace our current
approaches. The advantages of giving a one-time vaccination of a product not requiring refrigeration by a simple gene gun technique are obvi-
ous. Such a DNA vaccine could be taken further afield, allowing for
greatly expanding the populations covered. If more people were resistant to infection, the yearly epidemic of influenza would be broken.

Example—Hepatitis C Virus

Hepatitis C infection is the major cause of chronic liver disease and cirrhosis affecting up to 3% of all people. Acute infection with HCV progresses to a chronic infection in 80% of cases with nearly half progressing to liver failure. HCV (like HIV) mutates rapidly. A standard
vaccine to envelope proteins is therefore not very effective. At least six
different genotypes of HCV exist with varying geographic distribution:
1a is common in North America and Europe; 1b and 2 are common in
Asia. (The type 1 genotype is known to be associated with a poorer response to standard interferon therapy.)

HCV is a single-strand positive RNA virus, 9,500 nucleotides in
length, encoding a single open reading frame (ORF) that codes for several
structural and nonstructural proteins. Figure 4.3 demonstrates the major
features of the HCV genome. The ORF is translated into a long polypeptide that is then cleaved into multiple structural and functional proteins.
Think of the polypeptide as a template with a series of "punch-out"
pieces. This is a common viral genome strategy, different from the much
more complex structure of human genes. The translation of the HCV
genome is under the control of a structure called IRES (internal ribosome
entry site) that is upstream on the 5' end of the genome (Fig. 4.3). Our
knowledge of the genome is the basis for multiple new approaches
to therapy.

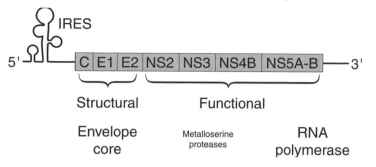

Figure 4.3. The genome of the hepatitis C virus has typical structures including regulatory elements [internal ribosome entry site (IRES)], and structural and functional genes. (See also, Wang and Heinz, 2000.)

The strategies for molecular treatment of HCV include nearly all of the modalities that we have considered. A vaccine for treatment or prevention of HCV must overcome the mutability of the virus. DNA vaccines that present antigen intracellularly to elicit a CD8 cytotoxic lymphocyte response are promising, with a number already in early clinical trials. DNA vaccines may need to be administered with adjuvant cytokines to boost the immune response. It is possible that vaccine-induced immune response will still need to be augmented with other antiviral treatment.

Beyond vaccines, anti-HCV agents under study include antisense RNA and ribozymes. The IRES region of the HCV genome is an attractive target for these agents. This region of the genome is similar (highly conserved) between the various genotypes and relatively immutable. The IRES serves to initiate all transcription of the viral genome. Liposome enclosed packets that are taken up by the cell can deliver antisense RNA against this structure. Alternatively, an adenovirus or other vector can deliver an engineered piece of DNA into the cell for transcription into an antisense RNA. A ribozyme that cleaves a critical portion of the IRES is another attack.

Beyond attacking the genomic structure of HCV, we can look at inhibiting its functional proteins. Like HIV, drugs that inhibit the HCV proteases or RNA polymerase can lead to long-term containment of the virus. These agents will likely work best in combination with vaccines or RNA antisense and ribozymes.

The search for molecular therapy of HCV parallels the treatment of HIV and many other viruses. This is good. We are finding common threads in molecular therapy; know the genome of the infectious agent and you will know where to attack it.

Aside: Infectious Vaccines

Imagine the following scenario. Sometime in 2002, a genetically engineered DNA vaccine for HIV is available. A United States–led international consortium of pharmaceutical companies announces that it can produce the vaccine for $20 a dose. It will distribute sterile single-dose injection units at cost on a priority basis, beginning with service to the countries that participated in the research and investment leading to the vaccine. The consortium anticipates production of 50 million doses the first year. The leader of an African nation with a population of 120 million people, 20% of whom are infected with HIV, announces that this is unacceptable. This leader has recently come to power in a riotous revolution, in part a reaction to that country's health crisis. The health minister of this country, a highly respected physician, amplifies his president's comments by announcing that his country will soon begin release of an infectious version of the HIV vaccine. An infectious vaccine is a microbe designed to spread from person to person to induce immunity to a specific antigen. Working in a World Health Organization (WHO) laboratory that was taken over during the recent revolution, he and his colleagues have inserted DNA that induces T-lymphocyte immunity to HIV into an influenza virus. The infectious virus carrying the DNA vaccine will soon be introduced into refugee camps located at this country's borders. These refugees will then be released to spread influenza and the DNA vaccine against HIV around the region. This, the health minister announces, is the only affordable and practical means of introducing the vaccine into the 120 million people of his nation. He also offers the modified influenza virus infectious vaccine to the health departments of other Third World countries.

The scientific problems with the scenario outlined above are great, especially the risk of mutation in the infectious virus rendering it virulent instead of protective. Beyond the scientific problems loom far greater social and ethical issues. As the technical capabilities of molecular biology increase and become widespread, so do both the benefits and risks of its products. An infectious vaccine is very close in technical specifications to an infectious agent designed to spread disease—an act of biological terrorism!

Many, if not most, people acquire immunity to viruses through subclinical or mild infections. In this sense, many people are immunized by an infectious "vaccine." The delivery of vaccines to large numbers of people is a complex and expensive task. The use of an infectious vaccine theoretically allows for inexpensive and rapid delivery of the antigen to a large number of people.

Another approach, long in development but now bearing fruit (pun intended) are edible vaccines (Langridge, 2000). The hepatitis B banana or the rotavirus potato are the goals of this approach. Transgenic plants that express bioengineered microbial antigens in their edible parts is another, safer method of vaccination. The plant can be grown locally, distributed like produce, and eaten even in the most primitive places, just where vaccination is most needed.

Genetic Susceptibility to Infection

Why do some people get sick when others do not, after apparently identical exposures to an infectious pathogen? The microbe may be the same, but the host is not. Host factors are complex. They cannot be studied, like bacteria, in a Petri dish. Molecular technology, however, has opened the door to the study of genetic susceptibility to infection, especially now that we have both the human and microbial genomes in hand. Susceptibility to infection is undoubtedly multifactorial with genetics being only part of it. Nevertheless, we can just begin to appreciate individual polygenic traits that allow specific microorganisms to gain an advantage. Table 4.4 lists just a few examples of genetic polymorphisms that predispose to infection. This is a field that is just beginning now that we have the tools.

Prions

Prions are proteins that have the capability to change the shape and function of other proteins. If that was all they were, we would call them enzymes. The changes caused by prions result in insidious neurologic diseases. We could therefore consider prions to be toxins. But prions also have the extraordinary property of making copies of themselves out of other proteins; they reproduce. Because they can reproduce, most people consider prions to be infectious agents. Prions are the only infectious pathogens known that do not contain either DNA or RNA. The prion protein (PrP^C) is encoded by a chromosomal gene. This normal protein when in proximity to the disease-causing protein designated (PrP^{Sc}) is changed into the disease-causing form. Both proteins have the same amino acid sequence, and the abnormal shape of PrP^{Sc} results from a different folding of the molecule. This folding of PrP^{Sc} can take more than one shape, and each shape is associated with a different clinical manifestation of the prion-based disease.

Sporadic Creutzfeldt-Jakob disease is the best known and most common from of prion disease, and is caused by an acquired somatic mutation in the prion protein gene (*PRNP*). The less common familial form of Creutzfeldt-Jakob disease is due to an inherited germline mutation of *PRNP*. Of special concern are cases of prion disease caused not by mutation but by exposure to the abnormal protein PrP^{Sc}. This protein is very resistant to heat and disinfectants. The infective form of the protein must reach the central nervous system where PrP^{Sc} can come into con-

Table 4.4. Genetic polymorphisms predisposing to infection.

Interferon-γ receptor mutations lead to fatal infections with usually
   nonpathogenic mycobacteria
Tumor necrosis factor polymorphisms are associated with susceptibility
   to malaria and other infections
Red blood cell hemoglobin and enzyme mutations are associated with
   relative resistance to malaria, e.g., sickle cell disease
Chemokine receptor polymorphisms allow HIV-infected individuals to delay
   or avoid progression to symptomatic illness
Natural resistance–associated macrophage protein 1 affects infection with tuberculosis
Mannose-binding lectin polymorphisms alter susceptibility to meningococcal disease

From Kwiatkowski (2000).

tact with the normal PrP^C. Mad cow disease in humans may be a new
variant of prion-based disease in which bovine prion protein is eaten
and then reaches the brain and begins converting good prions to bad.

Aside: What Is the Definition of Life in the Genomic Era?

Let's consider the issue of whether prions are alive as a test of the much
broader question of the definition of life in the context of our advanced
knowledge of the genome. A dictionary definition of life from the pregenomic era is "the condition that distinguishes animals and plants from
inorganic objects and dead organisms, being manifested by growth through
metabolism, reproduction, and the power of adaptation to environment
through changes originating internally." [1] Metabolism is the conversion of
carbohydrates to carbon dioxide and water. This definition is OK for large
plants and animals but is far from watertight. A candle possibly fits this
definition while a virus does not. In the twenty-first century, we add the requirement of genome. The blueprint of the organism is encoded in DNA or
RNA, and copies of that blueprint are shared with all offspring. We must
also consider deleting the requirement of metabolism. Many microbes
have very limited or unusual ways of satisfying their energy requirements.

Prions challenge any current definition of "life," and many people
do not consider them a living infectious agent, just infectious. Prions
make copies of themselves by changing the shape of other prions into
their own disease-causing form. Prions do not have a genome. They do
encode information in the way their protein is folded, but this is not
something we currently recognize as genomic.

Alu sequences, which we encountered in Chapter 1 as a form of
repetitive sequence in the junk DNA of the human genome, are another
test of the definition of life. Alu sequences, if alive, live only within the
genome of animals. As far as we know, they never encode a protein,

---

[1] The Random House Dictionary of the English Language, 1973, p. 827.

never metabolize. Yet they do reproduce and they do contain DNA. The number of Alu sequences in the human genome, already over 1 million, is slowly increasing with time.

Our understanding of life at the molecular level is far from complete. We must at the moment admit that we cannot write a watertight definition of what is alive and what is not.

Biological Terrorism

The threat of biological terrorism has deservedly received much attention. Molecular medicine adds both to the threat and to the defense against biological terrorism. On the threat side, it is not difficult to engineer microbes with enhanced virulence. We could, for example, transfect cobra venom into *E. coli*. We have seen enough gene diagrams already in this book to realize how easy it is to mix and match genes to create a "superbug." However, to paraphrase one expert, with anthrax or plague, what need do we have for increased virulence? Therefore, most of the recent concern about biological terrorism centers around how much easier it is to disperse organisms and how fast an infection can spread via global air travel. The bioterrorism events of October 2001 show that even standard mail can be a vehicle for the durable spores of anthrax.

Fortunately, molecular medicine has much to offer in terms of defense. Very rapid identification of an unexpected infection is key. As molecular diagnostics progresses, this becomes possible—including field scanners that detect infectious agents via DNA or protein markers. Rapid preparation of vaccines or other specific measures is the next step. The section Aside: Infectious Vaccines (see above) can prompt us to consider the possibility of counter-bioterrorism steps. For a view of the current state of bioterrorism preparedness, consult the Web site *www.bt.cdc.gov.*

Summary

Microorganisms have shown a great ability to live anywhere and mutate rapidly to adapt to changing conditions. They are formidable pathogens and the cause of a large fraction of all human diseases. Their genomes are very different from ours; this may be a great advantage when we begin to apply molecular therapies to infectious disease. We have a host of potential new ways of blocking the reproduction of microbes, including antisense, ribozymes, engineered antibodies, and vaccines. The threat of emerging infectious diseases has never been greater as the population of this planet moves further and faster, traveling beyond its natural host immunity. We need the new methods of molecular medicine to combat this threat.

Bibliography

Bunnell BA, Morgan RA. Gene therapy for infectious diseases. *Clin Microbio Rev* 1998;11:42–56.

Chen Z, Kadowaki S, Hagiwara Y, et al. Protection against influenza B virus infection by immunization with DNA vaccines. *Vaccine* 2001;19:1446–1455.

Kwiatkowski D. Genetic dissection of the molecular pathogenesis of severe infection. *Intensive Care Med* 2000;26 (suppl 1):S89–97.

Langridge WHR. Edible vaccines. *Sci Am* 2000;283:66–71.

Lauer GM, Walker BD. Hepatitis C virus infection. *N Engl J Med* 2001;345:41–52.

Prusiner SB. Shattuck lecture—neurodegenerative diseases and prions. *N Engl J Med* 2001;344:1516–1527.

Soini H, Musser JM. Molecular diagnosis of mycobacteria. *Clin Chem* 2001;47:809–814.

Wang QM, Heinz BA. Recent advances in prevention and treatment of hepatitis C virus infections. *Prog Drug Res* 2000;55:1–32.

www.cdc.gov/ncidod/eid/index.htm (CDC's journal *Emerging Infectious Diseases*).

5   Inherited Genetic Diseases

Overview

Initially, we might think that the genetics of inherited diseases is simple. We have decoded the genome. We have the technology to find the errors that lead to inherited diseases. What's the problem? Looking at a few examples, we will quickly see that our new molecular knowledge and technology lead to very great challenges in applications. To begin, we can consider screening for inherited disease. It sounds simple in principle; build the right DNA chips and collect blood from everyone. We will look at three straightforward inherited diseases—factor V Leiden–associated thrombosis, hemochromatosis, and cystic fibrosis—and then consider two complex polygenic diseases—atherosclerosis and diabetes mellitus. These examples will help our thinking to evolve from a simple view of the genetic basis of disease to an appreciation of complex interactions. The complexity, although daunting, offers many new "hooks" for molecular therapy.

Individual Genetic Screening for Inherited Diseases and Risk Factors

Let's make the assumption that technology will soon exist to probe the genome of an individual patient with accuracy and at a reasonable cost. What would a physician want to know about the individual nuances of the genome of a new patient? Certainly we would want to screen for errors of metabolism such as phenylketonuria (PKU), which, if not compensated for, can seriously damage a newborn. We now screen for PKU using a bacterial inhibition assay on dried blood spots. The test is sensitive, but not at all specific. Most infants (about 95%) who are positive on the screen will be negative on follow-up testing. Nevertheless, mandatory PKU screening has dramatically decreased the prevalence of mental retardation due to this disease. Many states test for several inherited diseases from the same dried blood spot required for PKU screening.

Molecular technology in theory can just as easily test for a hundred or even a thousand mutations as for one. Our knowledge of the human genome and our technology makes this possible. The ultimate would be to sequence the entire genome of every newborn, put the data in the medical record, and examine it as appropriate: pediatric diseases first, predisposition to chronic disease later.

We probably should consider what our first steps would be before we jump to the end point of a complete genome for every person. Table 5.1 lists some example diseases, all of them subject to debate, that might be candidates for our first screening arrays. All of these diseases can be subjected to genetic analysis now. What would we do if faced with DNA arrays that screened for a large number of inherited diseases and risk factors? I am confident that every reader of this book will soon be faced with such a problem. The whole purpose of our study of molecular medicine is to prepare us for this kind of challenge. The problems with expanded genetic screening for inherited diseases are immense. Each disease requires clinical knowledge and experience in order to interrupt the result and to advise the patient.

Simple Genetic Diseases

As examples, let us consider three specific inherited diseases: thrombosis due to factor V Leiden, hemochromatosis, and cystic fibrosis. Each of these is a relatively simple genetic disease, in that only a few mutations are responsible. Yet, we will see the complexity of understanding and using genetic information build as we progress through these three diseases.

Thrombosis and Factor V Leiden

Venous thrombosis leading to life-threatening pulmonary embolism is a multifactorial disease, but we now know that one significant factor is genetic. An inherited mutation in one or more of the genes that interact with the anticoagulant proteins (antithrombin or proteins C and S) is seen in about 20% of people with thrombosis. The factor V Leiden mutation, named for its discovery by researchers in Leiden, Holland, is the most common cause of genetic susceptibility to thrombosis (database links OMIM 227400; Locus ID 2153). This point mutation is a substitution of A for G at nucleotide 1691 causing a replacement of arginine [R] by glutamine [Q] at amino acid codon 506. The factor V mutation may alternatively be called $FVR^{506}Q$ or $FV:Q^{506}$ or R506Q or G1691A. Our terminology for gene mutations is still developing, and will certainly need a lot of help with inconsistencies by the time we have identified tens of

Table 5.1. Examples of genetic screening of inherited diseases.

Alcohol dehydrogenase polymorphism
α-1-Antitrypsin deficiency
Attention deficit disorder
Congenital hypothyroidism
Cystic fibrosis
Deafness
Familial adenomatous polyposis
Fragile X mental retardation
Galactosemia
HDL/LDL/VLDL receptor polymorphisms
Hereditary fructose intolerance
Hereditary hemochromatosis
Hereditary nonpolyposis colon cancer
Hereditary predisposition to breast and ovarian cancer (*BRCA1* and *BRCA2*)
Homocysteinuria
Huntington's disease
Insulin resistance
Lipoproteins, apoB, apoE polymorphisms
Long QT syndrome
Phenylketonuria (PKU)
Schizophrenia
Sickle cell anemia
Tay-Sachs disease
Thalassemia

thousands of specific mutations. The mutant factor V protein produced by this mutation is resistant to the action of activated protein C, thus breaking the feedback control of the anticoagulant branch of the coagulation pathway. Factor V Leiden is an autosomal-dominant mutation; heterozygotes have a 5- to 10-fold risk of thrombosis, homozygotes a 50- to 100-fold risk. Compared to other mutations that we will talk about in this chapter, factor V Leiden is straightforward; there are no alternate alleles. This is just a simple you-have-it-or-you-don't point mutation.

The factor V Leiden mutation is easy to test for. The mutation is diagrammed in Fig. 5.1. The wild-type allele is shown on top. A cut site for the restriction enzyme MnlI occurs at codon 506. In the R506Q mutation with an A instead of a G, the cut site is lost (as shown on the bottom). Figure 5.2 shows a traditional technology for factor V Leiden mutation testing, using polymerase chain reaction (PCR) amplification of a region around the point mutation. A gel electrophoresis of the PCR products digested with MnlI shows two bands in subjects heterozygous for the factor V Leiden mutation. One band of 163 base pair (bp) is from the chromosome with no mutation (wild-type allele); another band of 200 bp is from the mutated allele with a loss of the MnlI site. Homozygotes will show only one 200-bp band; subjects with no mutation will show only one 163-bp band. The technique illustrated in Fig. 5.2 is called restriction

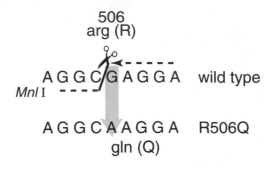

Figure 5.1. A diagram of the factor V Leiden mutation demonstrates a single base substitution that results in the loss of an MnlI restriction enzyme cut site.

fragment length polymorphism (RFLP) analysis. Recall that in Chapter 1 we carried out an analogous process to find an error in the preamble to the Constitution.

The lab test for factor V Leiden is simple; the medical problem is, How do we use this knowledge? As molecular technology progresses it will be simple and even inexpensive to test everyone for this mutation (as suggested in Table 5.1). Factor V Leiden mutation occurs in 5% to 10% of whites of European descent, but it is rare in Africans and Asians. I think that as of now, many specialists would recommend testing for factor V Leiden in white patients with an unexplained or unexpected thrombotic episode. What about testing for patients about to undergo hip surgery or other procedures that create a situation that predisposes to thrombosis?

The answer to the question of whom and when to test for a mutation is based on what additional medical steps will be taken if the genetic test comes back positive. This is a recurring theme that we will face for every genetic test that we discuss. Molecular medicine is mostly just medicine, a multifactorial scientific and judgmental practice that requires experience. A test for factor V Leiden mutation, if positive, could result in very positive prophylactic steps such as anticoagulation and closer monitoring in patients at risk for thrombosis.

Hemochromatosis

Hemochromatosis is a common, underdiagnosed disease whose worst manifestations can be curtailed by early treatment. The incidence of hereditary hemochromatosis is about 0.5% in the United States. Traditional testing for hemochromatosis begins with a serum transferrin saturation level. If this is greater than 60%, follow-up includes serum ferritin, liver function tests, and possible liver biopsy for quantitative iron and determination of cirrhosis. A DNA test for the *HFE* (OMIM 602421; Locus ID 1080) gene mutations responsible for hereditary hemochro-

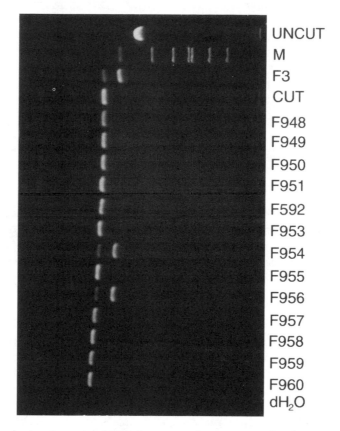

Figure 5.2. An electrophoresis of PCR fragments demonstrates the detection of the factor V Leiden mutation, using a polymorphism for MnlI. F3 is a heterozygote positive control, and Cut is a negative control. Two subjects show dual bands indicating heterozygosity. (Reproduced with permission from Friedline et al., 2001.)

matosis is now available, and in my opinion is a preferred screening method. A national patient advocacy group, American Hemochromato sis (*www.americanhs.org*), has argued for widespread genetic screening. This is in contrast, as we will see in the next section, to the Cystic Fibrosis Foundation advocacy group, which is cautionary about genetic screening for that disease.

The purpose of screening for hemochromatosis is to detect the disease before iron overload has caused symptoms. Buildup of storage iron in the liver causes dysfunction and eventual cirrhosis. In the pancreas, iron overload causes diabetes. The genetic test recognizes two mutations, only one of which is likely to cause disease. Table 5.2 lists the possible genotypes: wt/wt (normal or wild type on both chromosomes) versus a mixture of the C282Y or H63D mutations. Homozygotes for C282Y are very

Table 5.2. Genotypes of hereditary hemochromatosis (HH).

Genotype	Frequency in Caucasian population (%)	Frequency in HH patients (%)	Disease type
wt/wt	55–60	<0.1	None
wt/C282Y	13	2–3	Carrier; iron overload is rare
C282Y/C282Y	0.5	80–90	Iron overload highly likely
wt/H63D	25	1	None
H63D/H63D	2	3	Iron overload is rare
C282Y/H63D	2	3–5	Iron overload is rare, mild if it occurs

likely to have the disease. Heterozygotes are carriers. The incidence of the possible genotypes, both in the general population and in patients with the disease, is shown in Table 5.2. Genotyping as a screening method seems straightforward, and hereditary hemochromatosis may be one of the first diseases where genetic testing is itself tested. In 2001, a trial program of free genetic testing for hereditary hemochromatosis begins, with the first target area in western North Carolina.

Even with a simple test that identifies an important treatable disease, problems remain. Consider the pedigree for one family with hemochromatosis shown in Fig. 5.3. The dark square is Bob, our patient with the disease. We know that his mother and father must be carriers, but they show no sign of the disease. Bob's four children must either be carriers or possibly have the disease. They are too young to have clinical symptoms. Bob also has two sisters and a brother. As a doctor treating Bob, are you obligated to notify his brother and sisters to be tested, especially since you know that Frank, his brother, has been having some trouble with his liver? If you don't tell Frank, are you medically and ethically liable for letting a treatable condition become disabling? If you do tell Frank and he objects to this invasion of his privacy, are you liable for that? The lower panel of Fig. 5.3 is the same family pedigree after genetic testing. Frank (the second black square), as it turns out, does have the disease, but his and Bob's children are only carriers, and will not be affected. We are just beginning to learn how to handle genetic data. Not every family will want to be tested; not all families are geographically or emotionally close. Genetic counselors can tell us of many difficult issues, including surprises such as nonpaternity of fathers. As DNA testing for inherited disease becomes much more common, so will these issues.

Cystic Fibrosis

History of the Discovery of the Cystic Fibrosis Gene

The search for the cystic fibrosis (CF) gene is an intriguing story, and informative, since CF was one of the first inherited diseases dissected by gene mapping. In the late 1980s the story of the search for the CF

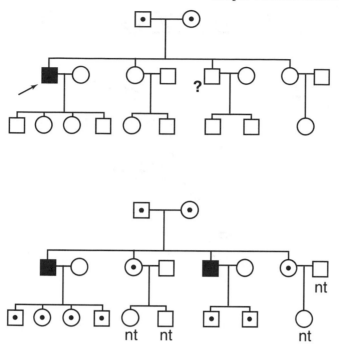

Figure 5.3. Pedigree of a family with hemochromatosis. The top panel shows a clinical pedigree. DNA testing on the family fills in many blanks as shown in the lower panel.

gene was news; each step was reported in the popular press, frequently ahead of publication in scientific journals. One of the main CF researchers, Francis Collins, is now head of the Human Genome Project, based on his experience and fame in the CF search. The story of the search for the CF gene caught the public's attention and introduced molecular medicine.[1]

Cystic fibrosis is a very common inherited disease. Research into the nature of CF has had extensive support from the Cystic Fibrosis Foundation (*www.cff.org*). Investigation into CF began with the cooperation of CF families and with the pulling together of a large number of pedigrees with blood samples. A large panel of then-known genetic markers was used to screen the DNA from these pedigrees. Soon, these data revealed that the gene was located on chromosome 7. By 1985, more linkage data localized the gene to a region on the long arm of chromosome 7 called 7q3.1. The gene was felt to be within a region spanned by two markers called met and J3.11. Figure 5.4 shows a diagram of chromosome 7 and the progressive steps taken in the late 1980s to localize the CF gene.

---

[1] Later, the O.J. Simpson trial would make everyone familiar with the uniqueness of their own DNA.

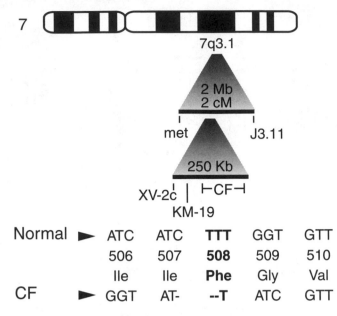

Figure 5.4. A diagram of chromosome 7 shows the most common mutation that results in cystic fibrosis (CF), a common autosomal-recessive disease.

There was quite a bit of excitement at the time. The region between met and J3.11 is less than 2 megabase pair (Mb), a genetic distance of about 2 centimorgan (cM). This meant two things. First, a genetic distance of 2 cM means that only 2% of the time will genes in this region be separated at meiosis. Genetic counselors could use studies of linkage to met and J3.11 to give a nearly definitive diagnosis of carrier status in families with CF.[2] Second, a physical distance of 2 Mb meant that researchers were getting close to the gene. Sequencing was very slow in the 1980s. To sequence 2 Mb might only be a few hours of work today, but back then it would involve several of the best labs for more than a year.

There was also an ethical dilemma about searching the last 2 Mb, not too different from current debates over the use of data from the Human Genome Project. The researchers who had spent years finding the general location of the CF gene were now being asked to give their probes and DNA to newcomers who might get most of the credit for doing the last piece of the search. Yet holding back on distributing scientific information and tools could certainly not be justified to CF families, who cared little about credit but wanted a cure. The search for the CF gene

---

[2]This may seem crude to us now when we have the entire gene in hand and the technology to detect specific mutations, but that was the way genetic testing began.

was like the California gold rush. Only in this scientific gold rush, the first prospectors who had done the long work in finding the gold fields were now being asked to give away maps and mining tools to the flood of new comers. History shows that the scientists involved followed a fairly altruistic course. They ran a fine line between announcing too much too soon, before it had been adequately tested, and being accused of holding too much back. Even so, several premature announcements of the discovery of the CF gene occurred between 1987 and 1989.

Finally, in 1989, the CF gene was found. Two laboratories, both of which had been major contributors to the accumulating research, announced that the mutation causing CF had been found. The mutation was named ΔF508. Figure 5.4 shows the DNA sequence in the normal genome and in the gene of a CF patient. The mutation is a deletion of three bases: one base in codon 507 and two bases in codon 508. This causes the deletion of a single amino acid, the phenylalanine residue normally present at the 508 position of the 1,480 amino acid chain of the CF gene.

Since the initial discovery, the genetics of cystic fibrosis has been found to be more complex than a single mutation, and we are still learning how to utilize DNA testing in this disease. The discovery and sequencing of the CF gene was a stellar success in the early years of molecular medicine. The DNA sequence when discovered told us about the structure of the protein and the mutations that made the protein nonfunctional. This information in turn gave us new targets for both drug and gene therapy.

Genetics of Cystic Fibrosis

Cystic fibrosis is a common autosomal-recessive inherited disorder in North American whites affecting approximately one in 2,500 infants. The disease results when a fetus receives two defective copies of the CF gene (*CFTR*) on chromosome 7 (OMIM 235200; Locus ID 3077). The cystic fibrosis gene codes for a protein called CFTR (CF transmembrane conductance regulator). CFTR acts as a chloride channel; disruption of its function leads to dehydration of secretions. Among the tissues most affected by loss of CFTR function are the bronchial and secretory cells of the airway, the exocrine pancreas, and the vas deferens. Pneumonia, especially due to *Pseudomonoas aeruginosa*, pancreatic dysfunction with malnutrition, and male sterility are the manifestations of cystic fibrosis.

Nearly 1,000 mutations have been mapped in patients with cystic fibrosis, but 70% to 85% are the ΔF508 mutation that was discovered first. All other mutations are infrequent; some are associated with a much attenuated form of the disease. A CF genotype has the potential to be predictive for each individual patient.

Prenatal diagnosis of CF is available to families with a history of the disease, or if the parents are known carriers. Figure 5.5 demonstrates

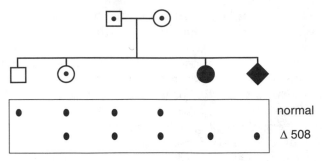

Figure 5.5. A pedigree demonstrates prenatal cystic fibrosis testing in a family with one affected child and a current pregnancy.

prenatal CF testing in a family that discovered after the birth of the third child that its members were carriers. The mother, father, and three children were studied along with fetal tissues obtained by chorionic villous biopsy from a pregnancy 20 weeks in progress. Note that both parents are carriers, a necessary condition for the fetus to have the disease. The oldest child is neither affected nor a carrier, the second child is a carrier, and the youngest child has CF. The diagnosis of CF in the third child is what led this family to seek genetic studies and counseling. The fetus of the current pregnancy was found to be affected.

About 80% of new cases of CF occur in families who do have a history of the disease. Should we screen for CF in all newborns? We thus would be better prepared for childhood complications such as pneumonia, and could potentially respond with early specific treatment. CF carriers would be identified before the age of child bearing. Couples would know if genetic counseling for CF might be an option for them. The negative aspects of screening are the stigma of identifying a person as a carrier. CF as a disease is treatable, possibly curable. Do we want to potentially alter behavior with a genetic screening test under these circumstances? All genetic screening tests bring us into a very controversial and uncharted area.

Knowledge of the cystic fibrosis gene immediately translates (pun intended) into more insight into the function of the protein. This information leads to new targets for drug therapy to overcome the defect in CF patients. Antibiotics, especially inhaled tobramycin, are directed against the airway infections. Other agents break up secretions (dornase, a DNAase) and improve ion transport (amiloride, a sodium channel blocker).

Gene therapy would be ideal, especially an inhaled vector that could replace the mutated CF gene in the airway. Despite many phase I trials, progress in gene therapy of CF has been slower than expected. Getting a good gene into the bronchial cell nucleus and then having it expressed is turning out to be difficult, as we saw in Chapter 3.

Polygenic Diseases

The three simple inherited diseases that we have just discussed are useful examples of molecular medicine. But most diseases are much more complex because they are both **polygenic** and multifactorial. We are not yet capable of understanding the interactions between many genes, even when we know of their existence. Yet, knowing something of the genetics of complex diseases opens new possibilities for treatment and prevention, even when our knowledge is incomplete. If we know that a disease has a genetic component, we can use the human genome database to find the genes. When we have the gene, we can hope to know its protein through our proteomics database. These data lead to knowledge about the basic biology of the disease. Knowledge is itself a most productive means of creating new therapies and strategies for disease prevention. Details about the proteins produced by the normal and mutated genes might, even in the absence of complete knowledge about the disease, lead to therapy.

We will consider the polygenic diseases atherosclerosis and diabetes mellitus. The message is that even without complete knowledge or discovery of all of the genes involved, molecular medicine can greatly advance treatment of these illnesses.

Atherosclerosis

Atherosclerosis produces its great burden of disease and death by compromising blood flow. Disruption of blood vessels with atheromatous plaque and superimposed clot are the end result of many factors. Lipid metabolism, coagulation cascades, platelet activation factors, arterial wall smooth muscle proliferation, and fluid dynamics are each in themselves complex fields. Atherosclerosis is the complex interaction of complex biology. The genetic portion of atherosclerosis is only slightly understood. We know of inherited disorders that, coupled with an atherogenic diet, predispose to early onset of this disease. How can we pick up one thread in a complex weave and hope to understand and intervene? Let us consider just one or two topics about atherosclerosis that may convince us that we can do just that.

Let's review a few of the features of the cholesterol metabolic pathway. Figure 5.6 shows the major features of cholesterol and triglyceride processing in the liver and blood. Triglyceride and cholesterol are excreted by a liver cell as a very low density lipoprotein (VLDL) particle with apoE and apoB proteins on the surface. Lipolysis of the particle releases triglyceride for uptake by fat and skeletal muscle cells. The remaining cholesterol and some triglyceride are repackaged as an intermediate-density lipoprotein (IDL) particle, still possessing apoE and apoB proteins. The IDL particle is taken up predominantly in recirculation through the liver, by binding to an LDL receptor. Some IDL particles are broken down into

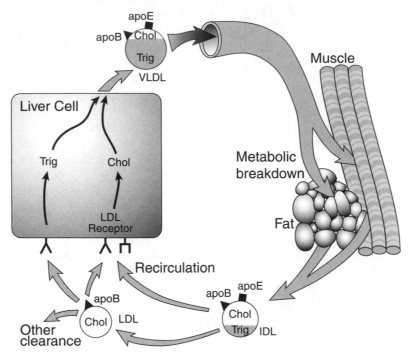

Figure 5.6. A schematic of cholesterol and triglyceride metabolism shows the recirculation of lipids and the many possible points of therapy.

nearly pure cholesterol containing LDL particles that have apoB on the surface. The LDL particles are also taken up by the liver and broken down at other sites.

ApoE

Let us consider one gene, *apoE,* within the multigene, multifactor equation for atherosclerosis. There are three alleles for *apoE* called E2, E3, and E4. This results in six possible genotypes: E3/E3, E3/E4, E3/E2, E2/E2, E2/E4, and E4/E4. The frequencies of those phenotypes in Caucasians are shown in Fig. 5.7. E3/E3, E3/E4, and E3/E2 account for 95% of individuals. Relative to E3/E3, the alternate genotype E3/E4 is associated with a 10- to 20-mg/dL increase in LDL, whereas E3/E2 is associated with a 10- to 20-mg/dL decrease in LDL. This is the genetic basis for about 10% of the LDL variation observed in a Caucasian population; the remaining 90% is due to multigenetic factors and environment. The E2/E2 genotype is seen in patients with familial type III hy-

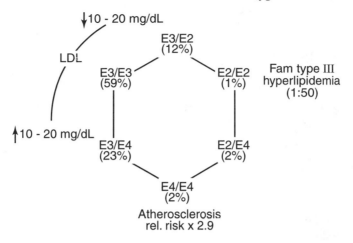

Figure 5.7. A summary of the alleles of *apoE* lists the incidence and associated risk.

perlipidemia. But only about 1 in 50 of individuals with the E2/E2 genotype manifest the disease (Breslow, 2000). E2/E2 may be a necessary precursor, but aging, obesity, and other environmental factors appear to be additional requirements for the genetic predisposition to type III hyperlipidemia to be expressed. The E4/E4 genotype has recently received the most attention. It is associated with a 2.9-times increased relative risk of coronary artery occlusion (Scuteri et al, 2001), as well as an increased risk of Alzheimer's disease and colon cancer.

This one *apoE* gene with only three alleles generates a picture rich in detail, and we certainly do not yet know the whole story. Looking at the broad outline of lipid metabolism, I can barely imagine how complicated the genetic and environmental factors must be. With so much complexity, and so much still unknown, what clinical value can molecular medicine offer? Well, we will have to learn; but many possibilities are presented. Complex lipid profiles can, when abnormal, be matched with partial genotyping to define more precisely individual patient risk and lipid altering therapies. The new field of pharmacogenetics is based on this kind of reasoning. An individual patient's genotype offers specific guidelines to individualized drug therapy. We will consider this further in Chapter 8.

Knockout Mice and Lipid Receptor Transplantation

We are not there yet, but a promising new molecular treatment of atherosclerosis is the transplantation of lipoprotein receptors. The knockout mouse, briefly introduced in Chapter 3, is a major tool of research in this area, and we should get to know it. A major way to determine the function of a gene is to create a mouse that lacks this gene, and see what

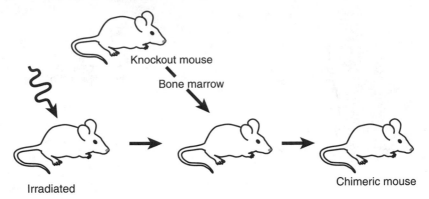

Figure 5.8. A diagram of an experiment using knockout mice to test the influence of LDL receptors.

goes wrong (or right). Knockout mice deficient in either low-density lipoprotein receptors (LDLr) or in apoE receptors have been engineered.

A set of clever experiments, performed in a number of laboratories, demonstrates the role of LDL and apoE receptors on macrophages. For years the foamy macrophage has been the "smoking gun" within the atheromatous plaque lesion. Now we are equipped to do experiments that re-create the scene of the crime, but remove the bullets from the smoking gun. Figure 5.8 schematically demonstrates a series of experiments in which bone marrow is transplanted from a knockout mouse into a normal recipient. The recipient mouse is first irradiated with enough x-rays to kill its own marrow. This mouse would die of bone marrow failure if not rescued with a bone marrow transplant. The transplant comes from a knockout mouse. What results is a genetically engineered chimeric mouse that is normal in all cells except for its blood cells that come from the donated marrow. The macrophages in the chimeric mouse are deficient in LDL receptors or apoE receptors, or whatever else we choose for our knockout mouse. We also need an experimental control group. Other mice are irradiated and undergo bone marrow transplant, only with normal mouse marrow instead of knockout marrow, just to be fair in creating as near-identical conditions as possible.

We then study the formation of atheromatous plaques in our chimeric mice versus control mice. They must all be fed an atherogenic diet to see the effect. Even in the laboratory atherosclerosis is a multifactorial disease. Mice with apoE-deficient macrophages have a marked increase in atherosclerosis (Fazio et al., 1997). Mice with macrophages deficient in LDL receptors have smaller atheromatous plaques than control mice. The control mice with intact LDL receptors have numerous foamy macrophages in their larger plaque lesions (Herijgers et al., 2000).

What can we conclude from these mouse studies? We have again experienced that atherosclerosis is a complex multigenic and multifactorial disease. We have also seen that changing the number and character of lipoprotein receptors can influence the rate of atheromatous plaque formation. The receptors influence the circulation of lipids within the blood. Receptors on macrophages also influence the rate of entry of lipid into plaque lesions. Currently, statin-like drugs are used to lower LDL and VLDL cholesterol. Transplanted macrophages with altered receptors may further limit the entrance of cholesterol into plaque. In Chapter 6, we will see that it is not too difficult to remove, genetically alter, and reinfuse bone marrow cells.

Atherosclerosis may depend on multiple genes and many environmental factors. Our understanding of some of this complexity and our ability to engineer drugs, genes, and cytokines and to manipulate the immune system give us many new attacks on the disease.

Diabetes Mellitus

Diabetes mellitus (DM) is a heterogeneous group of diseases resulting in hyperglycemia due to autoimmune destruction of the beta cells of the pancreas with no insulin secretion (type I) or to insulin resistance and altered insulin secretion (type II). Type II DM, the most common form, is multifactorial with both genetic and environmental risk factors. Identical twins (who also have a similar environment) have a 90% concordance for type II DM. A genetic component is a part of the equation; the disease is manifest when other factors such as obesity, lack of exercise, carbohydrate load, drugs, and hormonal status overload the body's glucose homeostasis.

Diabetes mellitus is most definitely not a simple genetic disease. For simplicity's sake, I would like to leave it out of this chapter on inherited genetic diseases. Yet molecular medicine is growing up; we cannot solve complex diseases such as DM, but we can shed light on the problem. In Chapter 2, we used the insulin gene as an example of one of the simplest and first cloned human genes. Human recombinant insulin was one of the first drugs of biotechnology. The gene encoding the insulin receptor is more complex. Surprisingly, genetic defects in the insulin signal pathway seem to be rare, and not a common cause of diabetes. There are rare special forms of diabetes mellitus, such as type I autoimmune DM and maturity-onset diabetes of youth (MODY), that do have a simple genetic basis or link. But the majority of cases of DM must be a complex interplay between many genes and developmental and environmental conditions. The current literature on diabetes speaks of a "thrifty phenotype." Low birth weight fetuses and infants of a certain genetic makeup utilize nutrition in a different way. If these children become exposed to food abundance and relative physical inactivity, they are highly prone to diabetes. The current epidemic of diabetes in Hispanic children in the United States seems to be an outcome of this pathway.

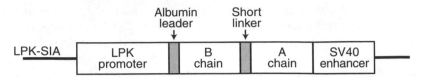

Figure 5.9. Genetically engineered insulin replacement gene. (From Lee et al, 2000.)

With diabetes mellitus, like atherosclerosis, what we can do when faced with complexity is to pick up a few threads of information. Let us consider genetic engineering and transplantation of the gene for insulin as a therapy for diabetes.

Insulin Gene Transplantation

We give diabetic patients insulin. Yet, careful monitoring of diet and blood glucose coupled with once or twice daily injections of insulin falls far short of the normal tight minute to minute homeostasis of glucose. A goal for better therapy has been to transplant a new source of insulin. Transplantation of beta cells of the pancreas has been problematic, with the usual need to overcome the body's rejection of foreign tissue. Transplantation of the insulin gene is an alternative. A transplanted gene must be put into the right environment to produce insulin in response to the correct signals. In Chapter 2 we discussed the factors that regulate insulin gene expression. These are summarized in Fig. 2.5. We must make a transplanted insulin gene respond in near-identical fashion; too much or too little insulin is life threatening.

A recent experimental model successfully transplanted an engineered version of the human insulin gene into a diabetic rat (Lee et al., 2000). This experiment demonstrates what must be achieved for an insulin gene transplant to work in humans. Recall Fig. 1.6, a map of the insulin gene. A single peptide proinsulin is transcribed from the *INS* gene. This proinsulin is converted into insulin by cleaving out the C piece joining the A and B chains. Proinsulin, unaltered, has only 2% of insulin's binding efficiency to the insulin receptor. Lee's group engineered a version of the INS gene that produced single chain insulin analogue (SIA) not requiring enzymatic cleavage. They did this by replacing the code for the C piece with a shorter code. The 35 amino acids (aa) in the C piece no longer need to be removed. Their SIA has 28% efficiency in binding to the insulin receptor and produces 40% to 50% of insulin's effect in producing a hypoglycemic effect.

So now they have a gene that does not require enzymatic processing. The next step is to make it respond to the signals that influence insulin. To achieve this they further engineered their gene to be under the control of the hepatocyte specific L-type pyruvate kinase (*LPK*) promoter.

This promoter responds to glucose like the promoters for insulin. The fully engineered *LPK-SIA* gene made by this group is shown schematically in Fig. 5.9. Compare this to the natural gene in Fig. 1.6.

The next step is to get this gene into the rat. These researchers used an established vector, the adeno-associated virus. Virus containing the *LPK-SIA* gene was infused into rats. The gene transfer cured rats that had been made diabetic with a chemical treatment, and the same curative effect was achieved in mice with autoimmune diabetes. The near-normal glucose levels of these animals persisted for 8 months (about 40% of a rodent life span). Simultaneous with the report of this success, another group achieved success with a different engineered insulin gene with a different promoter from the gut endocrine K cells (Cheung et al, 2000).

Ingenious, don't you think? Refine the design of a gene, give it new control elements, put it into a viral vector, and transplant it a diabetic animal. Our ability to manipulate genes is sophisticated. As we understand even bits of the puzzle in complex polygenic diseases, we have the possibility of intervening with this entirely new form of therapy.

Summary

We have the tools to screen individual patients for hundreds if not thousands of inherited diseases and risk factors. This new ability in genetic diagnosis far outstrips our experience with handling genetic data and with counseling. Some inherited diseases such as thrombosis due to a mutation in factor V Leiden or hereditary hemochromatosis are simple in that only one or two alleles are involved. Cystic fibrosis becomes more complex, with many alleles. Polygenic diseases such as atherosclerosis and diabetes are much more complex, the result of the action of many genes and the interplay of environment. Yet, we found that some knowledge leads to new therapies for these diseases. So even if we are not nearly done with learning about human genetics, we must begin.

Bibliography

Breslow JL. Genetics of lipoprotein abnormalities associated with coronary heart disease susceptibility. *Annu Rev Genet* 2000;34:233–254.

Cheung AT, Dayanandan B, Lewis JT, et al. Glucose-dependent insulin release from genetically engineered K cells. *Science* 2000;290:1959–1962.

Fazio S, Babaev VR, Murray AB, et al. Increased atherosclerosis in mice reconstituted with apolipoprotein E null macrophages. *Proc Natl Acad Sci USA* 1997;94:4647–4652.

Friedline JA, Ahmad E, Garcia D, et al. Combined factor V Leiden and prothrombin genotyping in patients presenting with thromboembolic episodes. *Arch Pathol Lab Med* 2001;125:105–111.

Gelehrter TD, Collins FS, Ginsburg D. *Principles of Medical Genetics* 2nd ed. Baltimore: Lippincott, Williams & Wilkins, 1998.

Goldstein JL, Brown MS. The cholesterol quartet. *Science* 2001;292:1310–1312.

Herijgers N, Van Eck M, Groot PH, Hoogerbrugge PM, Berkel TJ. Low density lipoprotein receptor of macrophage facilitates atherosclerotic lesion formation in C57BI/6 mice. *Arterioscler Thromb Vasc Biol* 2000; 20:1961–1967.

Lee HC, Kim S, Kim K, Shin H, Yoon J. Remission in models of type 1 diabetes by gene therapy using a single-chain insulin analogue. *Nature* 2000;408;483–488.

Lusis AJ. Atherosclerosis—insight review article. *Nature* 2000;407:233–241.

Moss RB. New approaches to cystic fibrosis. *Hosp Pract* 2001;36:25–37.

Press RD. Hemochromatosis: a "simple" genetic trait. *Hosp Pract* 1999;34:55–74.

Scuteri A, Bos AJG, Zonderman AB, Brant LJ, Lakatta EG, Fleg JL. Is the apoE4 allele an independent predictor of coronary events? *Am J Med* 2001;110:28–32.

So WY, Ng MCY, Lee SC, Sanke T, Lee HK, Chan JCN. Genetics of type 2 diabetes mellitus. *Hong Kong Med J* 2000;6:69–76.

6   Immune System and Blood Cells

Overview

The biology of the immune system and of blood cells is incredibly rich in detail. I prefer "rich in detail" as opposed to "complex." "Rich" invites us to explore the intricacies; "complex" is forbidding. Yet I must admit that the immune system is complex, probably as complex as the central nervous system. The immune system, like the brain, has input, output, central processing, memory, and schooling. Our understanding is far from complete, but molecular medicine has given us a much deeper look into the details and has provided many new tools for modifying the immune system. We will see that the diversity of the immune system in terms of its ability to respond to millions of antigens is derived from rearrangement of the family of genes that encode for the immune proteins—a process that, insofar as we know, is unique to the immune system. We also have discovered many of the cytokines and growth factors that modulate the immune system. Finally, we find that the major organ of the immune system, the bone marrow, can be easily removed, manipulated, and reinfused. Thus, the bone marrow has become a prime modality for genetic engineering to introduce genes that can correct metabolic and immunologic diseases.

Clonal Generation of Blood Cells

The generation of the major classes of cells that make up the immune system and blood cells is drawn schematically in Fig. 6.1. Each differentiated cell type is derived from clonal expansion of a hematopoietic stem cell. Under the appropriate stimuli, a stem cell undergoes a series of cell divisions with progressive amplification of the number of cells and differentiation of function, producing tens of thousands of mature progeny. Some of the steps in the path from stem cell to mature blood cell are irreversible. Other steps are modulated by growth factors, and

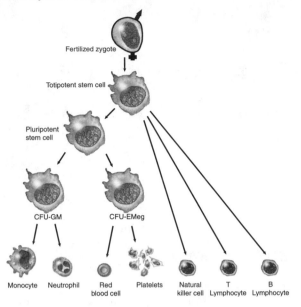

Fertilized zygote

Totipotent stem cell

Pluripotent
stem cell

CFU-GM   CFU-EMeg

Monocyte   Neutrophil   Red   Platelets   Natural   T   B
                        blood cell   killer cell Lymphocyte Lymphocyte

Figure 6.1. Stem cells generate all of the cell types of the blood and immune systems by a process of proliferation and maturation in response to growth signals. CFU-GM, colony-forming unit of the granulocyte/monocyte; EMeg, colony-forming unit of the erythrocyte/megakaryocyte.

the outcome may be varied according to the body's immediate need. In addition to the totipotent stem cell, which is the common precursor to all blood cell types, there are committed stem cells that can only produce progeny of a more restricted group of cells. The colony-forming unit of the granulocyte/monocyte (CFU-GM) is one such committed stem cell. While the bone marrow is the primary home of stem cells, they can also be found in blood and other organs. The totipotent stem cell of blood is not the same as an embryonic stem cell. It cannot be easily induced to differentiate into any tissue and is therefore not as good a source for cloning (see Chapters 3 and 8).

Lymphocytes are derived from marrow precursors, but then go on to spend their lives (which can be quite long) in lymph nodes and blood. Most of the functional maturation of lymphocytes occurs outside the bone marrow. Complex interaction between classes of lymphocytes in the node produces the final functional maturation of the many classes of lymphocytes necessary to maintain the complete immune system.

Molecular Biology of the Immune System

The immune system has a very complex set of tasks to carry out, interacting with every tissue in the body. We must envision a cellular ecology of the immune system to emphasize and appreciate the detailed in-

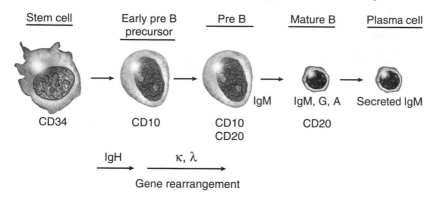

Gene rearrangement

Figure 6.2. The differentiation of B lymphocytes from the stem cell shows the progressive steps that include gene rearrangement of the immunoglobulin heavy chain (IgH) followed by κ and λ light chains. Nuclear and surface receptor proteins change as the cell matures, ending with surface and then secreted immunoglobulins.

teractions between cell types and growth factors. Figures 6.2 and 6.3 show just a few highlights of the development of some of the functional classes of lymphocytes. We can characterize the differentiation of lymphocytes by their surface markers, which are designated as CD numbers. Flow cytometry can immunophenotype a population of lymphocytes, measuring how many of each stage of differentiation are present.

B and T lymphocytes are derived from stem cell precursors in the marrow. Pre–B lymphocytes move to the lymph nodes and mature there. However, pre–T lymphocytes must first pass through the thymus, during fetal development. In the thymus, pre–T cells are schooled in proper function before going out into the blood or lymph nodes. This schooling includes among other things learning the fundamental distinction between self and nonself (foreign) antigens. Many mature B and T lymphocytes have undergone gene rearrangement to a fixed end state where they express a single antibody of unique specificity. These lymphocytes undergo clonal expansion, making many more copies of themselves when appropriately stimulated by an antigen or by interaction with another immune modulating cell. The members of the clone of lymphocytes are all the progeny of a single precursor. The clone produces the same unique antibody as the precursor.

Lymphocytes are capable of reacting to millions of potential antigens. The number of possible antibodies produced by lymphocytes is so great that the human genome does not have enough DNA to encode for all of the possible molecules. This was known before the era of molecular medicine. People wondered how the diversity of the immune response was generated. One theory was that the antibody molecule "folded" around an antigen and changed its structure in response to the antigen.

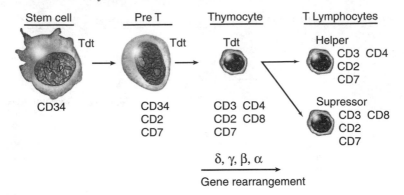

Figure 6.3. The differentiation of T lymphocytes also progresses from stem cell with early rearrangement of the T-cell receptor genes. Surface receptors (CDs 2, 7, 3, 4, and 8) characterize the functional state of the maturing T lymphocyte.

We now know that this theory is wrong. Gene rearrangement, as we will see, is the basis for the diversity in the immune system.

Mature differentiated lymphocytes can generate only copies of themselves. To generate a new clone with a different immune specificity, it is necessary to start over with a precursor cell and progress through the successive stages of development. The means by which exposure of the immune system to a new foreign antigen results in the production of a clone of antibody producing lymphocytes is very detailed. A few of the steps are shown in Fig. 6.4. Antigen must be processed and then presented to the immune system. This involves several classes of T and B lymphocytes as well as dendritic reticulum cells, monocytes, and other cells. All of these interactions involve either cytokines or cell surface receptors, and all of these are the products of genes. While the genes code for the tools of the immune system, the complexity is the result of "self arrangement." The function of the immune system depends on its history of exposure to antigens. Identical twins have different immune systems (as well as different central nervous systems) because of this.

Immunoglobulin Gene Rearrangement

Lymphocytes irreversibly rearrange their genome as they differentiate from stem cell to mature B or T cells (see Figs. 6.2 and 6.3). This phenomenon of immune gene rearrangement is the basis for the generation of diversity of antibody production within the immune system. Let's consider in detail how gene rearrangement occurs as B lymphocytes mature. An immunoglobulin molecule (IgG type) is composed of two heavy chains and two light chains. Each part of the heavy and light

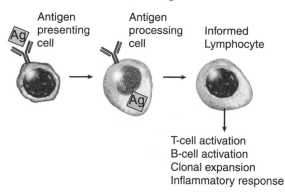

Figure 6.4. The functional interaction of lymphocytes is very complex. A greatly simplified schema shows that antigen is first presented to an immune cell, and then processed internally with subsequent instruction of a T lymphocyte. The interaction of these cells via surface receptors, secreted immunoglobulins, and cytokines leads to specific immune responses.

chains has a number of alternate possible genes that must be brought together to make a single complete protein molecule. This is diagrammed in Fig. 6.5. The genes that encode for the heavy chains are located on chromosome 14; those for the light chains are on chromosomes 2 and 22. It is necessary to select one gene from each of the possible alternate genes (which as a group constitute a gene family) and join them together to make a complete fusion gene for the protein. An analogy is to think of the immunoglobulin gene family as a deck of playing cards. Each mature lymphocyte constitutes one play of the hand where five or six cards are drawn from the deck and are laid out as the genotype for this cell. In this way, the immunoglobulin gene family consisting of tens of individual genes can be arranged into millions of possible genotypes, each capable of making a different antibody molecule.

The process of gene rearrangement for the heavy chain portion of the immunoglobulin molecule is shown schematically in Fig. 6.6. Genes must be joined from the V, D, and J families to produce a rearranged VDJ fusion gene that codes for the heavy chain. The VDJ piece will then be joined to the C or constant region. As a pre–B lymphocyte in the germinal center of a lymph node or the bone marrow differentiates, it first splices several of the six J segments to several of the 20-plus D segments. The genetic material between the splice is discarded! Next a few of the 50-plus V segments are joined to the DJ piece. Again the spliced out genetic material is discarded. The maturing lymphocyte can never reverse its destiny. It has permanently given up large parts of its immunoglobulin genes. It is now committed to producing only one specific antibody. The same gene rearrangement happens for the kappa

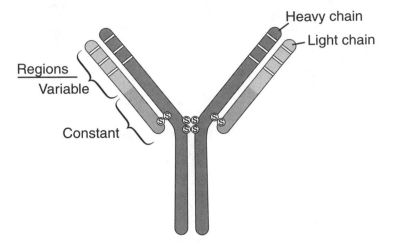

Figure 6.5. An immunoglobulin molecule is made up of two heavy chains and two light chains, joined together by multiple disulfide bonds (S-S). The variable and constant regions correspond to the V and Cμ genes (see Fig. 6.6).

light chain genes on chromosome 2 or for the lambda light chain genes on chromosome 22.

By now you might be thinking that this seems a little hard to pull off. It is. A lymphocyte succeeds in being its own genetic engineer only about one time in three tries. Normally, lymphocytes that fail in their attempt at gene rearrangement undergo apoptosis. Part of the reason for this low success rate is that it is necessary to link the genes in phase so that the recombined genetic sequences will be in an open reading frame (ORF) for translation of messenger RNA (mRNA) into a protein. Perhaps you can see the potential for trouble. All of this splicing activity opens the genome to the possibility of an error. In the next chapter we will see that malignant lymphoma is the result of the gene rearrangement process gone bad.

As far as we know, lymphocytes are the only cells in the body that rearrange their genes. No other cell has yet been discovered where genes are selectively spliced as part of a normal process. Lymphocytes rearrange their genes to achieve antigenic diversity. From only tens of genes, millions of antibodies can be produced. This gene rearrangement is so clever that I believe there will be other systems in the body where it will be discovered, such as the brain.

T-Cell Receptor Gene Rearrangement

The molecular biology of T lymphocytes is generally quite similar to B lymphocytes with the exception that the protein produced by gene rearrangement for T cells must stay fixed to the surface as a receptor mol-

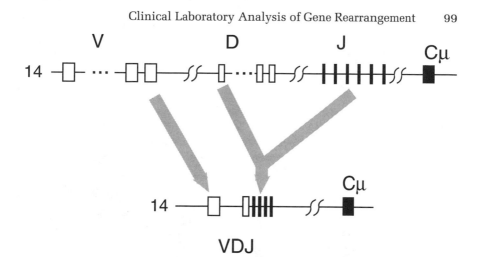

Figure 6.6. The heavy chain of immunoglobulin is formed by gene rearrangement and fusion of exons from multiple subfamilies on chromosome 14.

ecule. Furthermore, the T-cell receptor (TcR) acts only indirectly with foreign antigen rather than with the direct binding that is seen in B lymphocytes. The T-cell receptors that control immune interaction contain four major protein chains: α, β, γ, and δ. These four proteins are combined to generate the very complex surface receptor molecules. These proteins are synthesized from rearranged genomes in a process like the rearrangement of immunoglobulin genes in B lymphocytes. Figure 6.7 is a diagram of the T-cell receptor. It is a complex assembly of globular proteins (α, β, γ, δ, ε, and ξ), each one of which is made up of several polypeptide chains coded for by different rearranged genes! Figure 6.7 is just the CD3 "uncluttered" T-cell receptor; as the T lymphocyte matures toward helper or suppressor subsets, additional components are added to the complex.

Our knowledge of functional genomics is sorely tested and found to be incomplete at this point. We have stressed throughout this book how a genomic sequence codes for a protein. What we have not discussed is the additional folding, aggregating, and spatial arranging of proteins necessary to form a structure like the T-cell receptor. We don't know how this happens.

Clinical Laboratory Analysis of Gene Rearrangement

Leukemia and lymphoma are malignancies of the blood cells and immune system. They are characterized by a clonal expansion of an abnormal precursor cell. There are many ways to detect the presence of an abnormal clone of cells, and these methods are becoming important in the diagnosis of leukemia and lymphoma. Consider the normal distribution

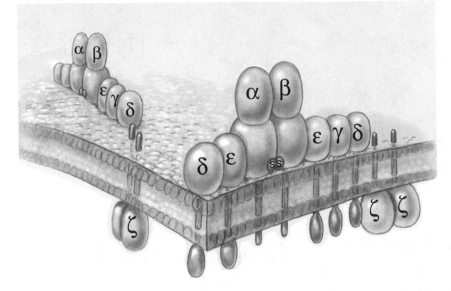

Figure 6.7. A diagram of the CD3 T-lymphocyte receptor demonstrates it to be a complex assembly of a number of proteins that connect the cell surface with the interior.

of rearranged genomes in a sample of lymphocytes taken from the blood or a lymph node. These lymphocytes will be polyclonal, consisting of a mixture of B and T lymphocytes that are primed to react to a wide variety of antigens. If the immunoglobulin gene is probed in a Southern blot or polymerase chain reaction (PCR) analysis, each different lymphocyte will give a different banding pattern from the rearranged immunoglobulin genes. A single broad band that is in reality a smear of thousands of faint bands overlaying each other will be seen. This is exactly the same result we get if we look at serum immunoglobulins by immunoelectrophoresis; the many different IgG molecules form a broad band.

In a lymphoma, many of the lymphocytes in a blood sample or lymph node biopsy are from one clone. Genetic analysis of the rearranged immunoglobulin genes will show one sharply defined band standing out from all the rest. Figure 3.2, presented when we first discussed the Southern blot method, is an example of an analysis of a T-cell lymphoma showing a dominant single clone of a rearranged T-cell receptor gene. Figure 6.8 shows a PCR analysis of rearrangement of the immunoglobulin gene. In this analysis, we are looking for a rearranged immunoglobulin gene that results from a t(14:18) chromosomal translocation. The nature of follicular lymphoma and its erroneous t(14:18) immunoglobulin gene rearrangement will be discussed in the next chapter. I present Figure 6.8 here to show how the analysis of immune gene rearrangements can lead to a detection of the abnormal clones that characterize lymphoma. We do

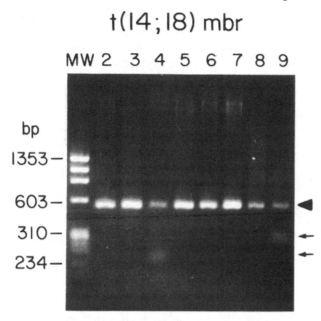

t(14;18) mbr

MW 2 3 4 5 6 7 8 9

bp

1353 —

603 — ◀

310 — ←

234 — ←

Figure 6.8. An agarose gel electrophoresis of the products of a polymerase chain reaction (PCR) to detect a clone of abnormal B lymphocytes associated with a t(14:18) chromosomal translocation. A positive result is seen in patient samples 4 and 9 (arrows). The large arrowhead marks the product of a normal gene, amplified as a control.

not know how often errors of immune gene rearrangement occur, nor do we know how often errors escape detection and progress to lymphoma. PCR is a very sensitive method for detecting lymphoma cells. Other methods include **Fluorescent in-situ hybridization (FISH)** and immunophenotype by flow cytometry. FISH is a good technique to search out very specific errors, especially those that result in chromosomal translocation. Immunophenotype looks at the protein surface receptors on lymphocytes; an abnormal pattern reveals the functional class or cell lineage of lymphoma. Analysis of gene activity of a lymphoma following therapy is possible using an expression microarray chip similar to the example demonstrated in Figure 3.6. All of these molecular methods have made lymphomas and leukemias the best-studied tumors.

Bone Marrow Transplantation

The transplantation of bone marrow allows for reconstituting the blood cells of the host. Bone marrow is unique among organ transplants in a number of ways. Surgery is not required. Bone marrow cells are harvested by aspiration from the donor and transplanted by intravenous infusion into the host in a manner identical to blood transfusion. The donor bone marrow stem cells that are transfused into the peripheral blood home in to the bone marrow, probably in response to local factors in the microenvironment within the marrow spaces of bones. The donor is only temporarily depleted of bone marrow tissues; regrowth is rapid. Donor stem cells can be taken from either aspiration of the bone marrow or from the blood by pheresis, concentrating the few stem cells. Another source of hematopoietic stem cells is cord blood, the discarded fetal blood that drains from the placenta after delivery of the newborn. Cord blood is rich in immunologically naive stem cells and can be transplanted into many hosts regardless of tissue match.

The bone marrow consists of many cell types, but its function can be completely regenerated by infusion of a small number of stem cells that are capable of dividing to increase their numbers. Under appropriate growth factor signals, bone marrow cells differentiate to produce all the red and white blood cell elements of the body. The bone marrow is an immunologically active organ in that it contains and regenerates lymphoid cells as well as granulocytes, monocytes, red cells, and platelets. Bone marrow transplantation offers the unique problem of graft-vs.-host disease (GVHD) because the donor bone marrow contains immunologically active cells that can fight back against the host tissues. This means that in addition to the standard problems of the donor rejecting the organ, bone marrow transplantation introduces the problem of the organ rejecting the donor! GVHD can be fatal.

Bone marrow transplantation is used to treat a number of clinical conditions (Table 6.1), including (1) hematologic malignancies, in which the malignant marrow is killed by intensive chemotherapy and a transplant is given; (2) serious congenital defects of hemoglobin synthesis, such as thalassemia major and sickle cell disease; (3) congenital severe combined immune deficiency (SCID), in which bone marrow transplantation reconstitutes the immune system; (4) metabolic diseases, such as Gaucher disease, in which improving bone marrow–derived phagocyte function can diminish the systemic manifestations of the illness; and (5) metastatic cancer, in which the bone marrow is killed by the toxic antitumor chemotherapy.

Molecular medicine has made possible diagnostic tests that aid bone marrow transplantation. Using DNA fingerprinting, it is possible to assay a blood sample and determine whether the cells are of host or donor origin. This test can be used to determine whether the engraftment has "taken." Sometimes after bone marrow transplantation, the marrow becomes chimeric, consisting of both host and donor stem cells. When bone marrow transplantation is used for the treatment of leukemia, DNA probes can monitor for the presence of a small amount of residual leukemia as well as detect engraftment from the donor cells.

Table 6.1. Candidate diseases for bone marrow—directed gene therapy.

Disease	Gene	Target cells
Adenosine deaminase (ADA) deficiency	Adenosine deaminase	Lymphocytes
Sickle cell anemia	Hemoglobin A	Bone marrow erythroid cells
Hemophilia	Factor VIII or IX	Liver macrophages
Gaucher's disease	Glucocerebrosidase	Liver macrophages
Chemotherapy-related myelosuppression	MDR1	Bone marrow stem cells

Bone marrow transplantation can be allogenic, from a human leukocyte antigen (HLA)-matched donor, usually a sibling; from cord blood, which is allogenic but too immunologically naive to recognize an HLA mismatch; or autologous, meaning that the marrow is taken from the patient, stored and treated outside the body, and then reinfused back into the patient. Autologous transplants are useful in permitting the intensive chemotherapy necessary to treat metastatic cancer. Bone marrow cells are harvested, kept outside the body during the infusion of chemotherapeutic drugs, and then reinfused.

Genetic Engineering of Bone Marrow Cells

The bone marrow is an enticing target for attempts to introduce genetically engineered cells into body because it is such a large factory for cell production. Genetic engineering of the bone marrow takes advantage of the relative ease of removing and reimplanting the cells. Bone marrow is also ideally suited to genetic engineering, since only a few bone marrow stem cells are capable of repopulating the entire bone marrow. In humans, reinfusion of 10^{10} marrow cells, approximately 10 g of tissue, results in repopulating the entire bone marrow, which is close to 1,000 to 2,000 g of tissue. It is theoretically possible that the entire human bone marrow could be reconstituted by engraftment of a single stem cell. Genetically engineered bone marrow stem cells independently seek out their correct location, traveling from an injection site to the bone marrow. Only a few cells can regrow into a large number and thus greatly amplify the effect of the transplanted gene.

The First Human Gene Transplant

The first human gene transplant was a momentous event, even though it was a small step, carried out quietly and anonymously at the National Institutes of Health (NIH) on September 14, 1990. For the first time humans were not only examining their genetic makeup but were attempting to rewrite it. The genes of all organisms are in a constant state of evolution. Over time, the genetic makeup of a species changes to adapt better to its environment. In 1990, only 15 years after the start of the recombinant DNA revolution and 10 years before the complete listing of the human genome, the first gene transplant was made into a patient in an attempt to alter the genetic makeup of the individual to correct an error. The ability of humans to read and reconstruct their own genetic makeup is unprecedented in the scheme of biological adaptation to the environment.

In Chapter 1, we used the analogy of the human genome compared to a library. One of the flaws in the analogy is that the human genome was not a library in that we cannot add to and alter its contents. With gene transplantation, this changes. Mankind now may read what is written in the genome, as well as add new information and make corrections. The power to alter the human genome is not without danger. We have only started on understanding the genome, and already we have seized the courage to consider changing it.

The first human gene transplant was done for purely medical reasons to help cure a child with a lethal disease. The events leading up to this human gene transplant were dramatic. The regulatory steps and review of the process by government agencies were more complex and took more time than did the genetic discoveries and manipulation of the gene. The disease affecting the patient was adenosine deaminase (ADA) deficiency—a genetic, inherited disease that creates a clinical state of SCID. It is the cause of approximately 25% of all cases of SCID. Patients with ADA deficiency begin to develop recurrent and severe opportunistic infections beginning in infancy. These infections, if untreated, are usually fatal within the first year of life. Even with intensive treatment of each subsequent infection, the condition is lethal. In the past, attempts have been made to place children with ADA deficiency into a sterile environment, a "boy in a bubble."

Adenosine deaminase deficiency is inherited as an autosomal-recessive condition. Fortunately, it is quite rare, with an estimated incidence of 10 per million births. The cause of the immune deficit in ADA deficiency is selective toxicity to lymphocytes, both T- and B-cell lineages, due to accumulation of metabolites of deoxyadenosine. Lymphocytes are hypersensitive to deoxyadenosine metabolites. Lymphocytes are the only cells severely damaged by ADA deficiency syndrome.

There are several reasons for choosing ADA deficiency as a target for gene transplant therapy. Drs. Anderson, Blaese, Rosenberg, and their collaborators at NIH wished to do a gene transplant that would not require altering the germline genetic makeup of an individual. That is, they did not want to correct the defect by making the person a transgenic organism. This is for reasons of safety and concern over altering the genetic makeup of a human. In theory ADA deficiency could be corrected by inserting a good copy of the gene into only a few blood lymphocytes, a first small step for gene transplantation.

A few patients with ADA deficiency have been cured by a successful allogenic bone marrow transplant from an HLA-matched donor. Regrettably, only small minorities of patients can be matched, which limits this therapy. The success, however, demonstrates the principle: get enough functional cells into the patient and the defect is corrected. The other reasons for selecting ADA deficiency for a human gene transplant are tactical: The human ADA gene had been cloned quite early. The ADA gene system is not part of a complex multigene process. The transplantation of genetically altered lymphocytes transfected with a normal gene for ADA should provide a source of enzyme. The overall level of expression of the transplanted ADA gene is not critical, since a wide range of levels of the enzyme corrects the defect without introducing any new toxicity problems. Furthermore, clinical improvement in immune function is seen with only a fraction of lymphocytes having corrected enzyme levels.

To begin the transplant, blood lymphocytes were removed from the patient and taken to the laboratory. There the cells were transfected with a correct copy of the ADA gene in a viral vector. One theoretical danger was that the viral vector, when the cells were reintroduced into the patient, might continue infecting other cells. This could cause the ADA transgene, made in the laboratory, to enter other cells besides the intended target, even other people. However, animal studies showed that there was no risk of unintended spread of the gene by the vector. After verifying that the patient's lymphocytes had taken up the gene in the laboratory, the actual transplant was begun. No news coverage was permitted. The genetically altered blood cells were simply infused back into the patient through an arm vein. Repeated infusions of genetically modified lymphocytes were continued for 2 years, in this and in another patient. These cells survived in the blood, and the levels of ADA were increased. The patients were kept on additional ADA therapies; no attempt was made to declare the transplant as the cure. It was a small first step, demonstrating feasibility and looking at many of the technical details of gene transplantation.

Aside: Chimeric Man

Human gene transplantation may progress very rapidly. Even as we learn the basic science and technology, we must also look further and see what possible outcomes may be possible. Consider the following. Shortly after conception, we transfect the fertilized zygote with genes offering lifelong immunity to a number of viruses, such as influenza, HIV, hepatitis C virus (HCV), and human papilloma virus (HPV). This would be in effect, in utero DNA vaccination. Later, at the moment of birth, we may wish to take a sample of cord blood. This would be a good time to add a few genes to the bone marrow of the newborn. We could for example, "improve" on lipid metabolism by lowering cholesterol. A short time later, we will screen the infant child for inherited genetic diseases and risk factors. We may choose some specific genetic therapy now, or wait until the appropriate time. As the child ages, we will add a few more DNA vaccines for infectious diseases, and a few more gene transplants to improve risk factors. As life progresses, and so does molecular medicine, many more gene therapies may be used. Stem cell transplant of cartilage to the knees may be indicated for the first signs of chronic damage due to high school football or skiing. Decades latter, our now middle-aged patient will receive more DNA-based therapies for specific illnesses. Medication for chronic diseases may be given as a transgene that manufactures the drug in the bone marrow or liver, removing the daily dosing of pills. Late in life, gene therapy will be aimed at treating cancers and providing regeneration for aging organs.

A **chimera** is an organism with a genetic mix of the original germline genome with additional genetic material derived from other means. Until very recently, when we acquired the ability to transplant genes, chimeras were quite rare and usually resulted from an organism having more than two parents by a genetic accident. In the future, mankind may choose to become chimeric by adding therapeutic genes to our natural genome.

Summary

The immune system is a complex interaction of many cell types distributed throughout the body and flowing in the blood. Molecular medicine has dissected this system further than any other because of its accessibility and the relative ease of growing bone marrow cells in vitro. We find that lymphocytes are their own genetic engineers, rearranging their genes as needed to respond to antigens. The interaction of the many classes of immune cells is modulated by cell surface receptors and secreted cytokines. We are learning the language of the immune system. Expression microarray analysis and other techniques show us what the cells are saying. It follows from this knowledge that the blood and bone marrow are chosen as the first cells to receive a transgenic modification in an attempt to treat a human disease.

Bibliography

LeBien TW. Fates of human B-cell precursors. *Blood* 2000;96:9–22.
Rich RR, ed. *Clinical Immunolog.* St. Louis: Mosby-Year Book, 1996.
Simpson S, Hurtley SM, Marx J, eds. Frontiers in cellular immunology. *Science* 2000; 290:79–100.

Stamatoyannopoulos G, Majerus PW, Perlmutter RM, Varmus H, eds. *The Molecular Basis of Blood Diseases*, 3rd ed. Philadelphia: WB Saunders, 2001:791–831.

Willis TG, Dyer MJS. The role of immunoglobulin translocations in the pathogenesis of B-cell malignancies. *Blood* 2000;96:808–822.

7   Cancer

Overview

Cancer is a disease of uncontrolled cell growth. Cancer results from a series of acquired defects in the DNA that cause deregulation of the cell's growth processes. The damaged cell transforms from benign to malignant and becomes independent of normal regulatory signals. This transformed cell multiplies into a clone of malignant cells, eventually developing into a tumor. The malignant tumor infiltrates adjacent tissues and metastasizes.

The molecular approach to cancer gets to the heart of the matter. Molecular studies have uncovered many of the details whereby a normal cell becomes cancerous. These discoveries have revealed cancer to be a multistep process, involving the progressive loss of control by the mutated cell with failure of DNA repair systems. We have discovered oncogenes that control cell growth through the cell signaling pathway. When these genes are mutated, regulation is lost. We have also discovered tumor suppressor genes that are the last line of defense in DNA repair. Cancer is always characterized by mutations in oncogenes and tumor suppressor genes.

We begin with a description of carcinogenesis at the molecular level. That knowledge is then applied to the clinical problems of colon cancer, cervical cancer, and lymphoma. Our newly gained knowledge of the molecular basis of cancer immediately suggests new molecular therapies.

Multistep Pathway of Carcinogenesis

Carcinogenesis is a process in which external agents (carcinogens) cause mutations in oncogenes and tumor suppressor genes leading to cancer. Chemicals and radiation are the main classes of carcinogens. Carcinogens damage DNA by causing strand breaks or interfering with

DNA replication, either of which can cause a mutation.[1] Damage to DNA may also arise during DNA replication by errors in copying. Most damage to DNA is repaired. If the DNA cannot be repaired, the cell generates an internal signal that leads to apoptosis. Some viruses suppress DNA repair, and are another possible source of mutations. Even when mutations accumulate in DNA, most of the damage occurs in silent locations within non dividing cells of various somatic tissues. Only when DNA damage interferes with the normal pattern of cell growth, and this damage is not repaired, do we get cancer. Thus cancer is a disease of multiple etiologies due to errors in the processing of information coded by DNA. Every living human cell is required to be both an individual entity and a loyal citizen of a multicellular organism. Cancer is a breakdown in the interdependence of cells.

Multiple steps are necessary for a fully invasive malignancy to develop. Figure 7.1 schematically demonstrates this process as a series of multiple hits. When a normal cell suffers a "hit" damaging its DNA, it may die directly from that damage. Alternatively, it may die a programmed cell death by apoptosis as part of the DNA repair process. Rarely, the damaged cell may survive, unrepaired and dysplastic, perhaps with a growth advantage. Among the clonal offspring of this damaged cell, a second hit occurs with much the same range of consequences. A third hit deepens the damage. Either the cell dies or eventually a frankly malignant cell is created. Prior mutations in this cell may now be expressed increasing its virulence. These events take time. Cancer is a disease that increases dramatically with age. Figure 7.2, compiled from a large body of cancer statistics (see Ries et al., 1996), shows the exponential increase in cancer mortality beginning at age 50.

Oncogenes

Oncogenes are growth control genes present in the human genome as well as in most complex organisms. The name "oncogene" was give to this group of genes because they were discovered initially in cancer cells and in viruses associated with cancer. Years ago, we thought that perhaps cancer was caused by catching an oncogene from a virus infecting a human cell. We later learned that viruses actually caught the gene from us. Retroviruses contain a special form of oncogenes in their genome that they have borrowed from cells that they infect. They use the oncogene as a key to unlock the cell's growth regulation to make the cell a more favorable host for the virus. Only rarely does the virus initiate all the steps necessary for a tumor.

---

[1]An acquired mutation in a somatic cell during carcinogenesis is very different from an inherited mutant allele. In carcinogenesis, a mutation is sporatic damage to DNA that changes or stops the function of a gene.

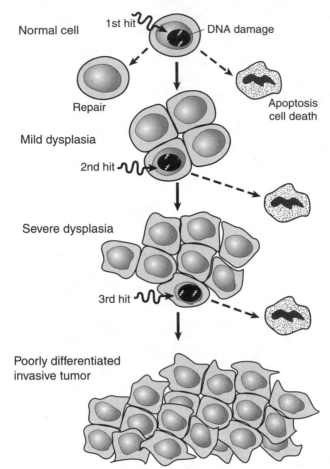

Figure 7.1. The multiple steps of carcinogenesis begin when a normal cell suffers a "hit," damaging its DNA. The cell may die from that damage or may undergo apoptosis if the damage cannot be repaired. Rarely it will survive unrepaired. A second hit to its clonal offspring continues the process. If the cell suffers yet a third hit, and fails to undergo apoptosis for a third time, the result will be an invasive tumor.

In their normal unmutated form, oncogenes direct physiologic functions in the cell signaling pathway (see Chapter 2 and Fig. 2.6). Table 7.1 gives a few examples of oncogenes, including their role in the cell signaling pathway and tumors associated with mutated forms of the oncogene. There are over 100 oncogenes discovered so far.

The cell signaling pathway is crucial to maintaining regulation of individual cells in a multicellular organism. Cancer cells act independent of the organism, failing to limit their growth in response to the ap-

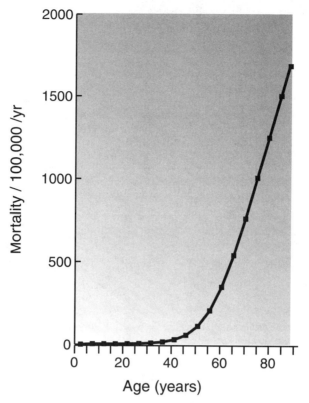

Figure 7.2. The death rate from cancer rises dramatically with advancing age.

propriate signals. They also fail to carry out their normal cellular function, and they are not limited to remaining in the place of origin. The clinical disease cancer results from all these failed steps in cell behavior. Tumors grow and then invade and metastasize.

ABL

ABL is an oncogene that encodes a tyrosine kinase. This tyrosine kinase serves as a cell surface receptor in the cell signaling pathway (recall Fig. 2.6). Mutations in tyrosine kinases cause the receptor to malfunction, deregulating the cell's response to external signals. Shortly we will see how an *ABL* mutation "causes" chronic myeloid leukemia. *ABL* is a classic oncogene having all of the characteristics defined in Table 7.2. It is a normal gene required for normal cell functioning. It also exists in a viral form v-*ABL* in the Abelson murine leukemia retrovirus. In cancer, *ABL* is an acquired (not inherited) somatic mutation that results in a hyperfunctional state. Tyrosine kinases in general are important targets for

Table 7.1. Examples of a few different types of oncogenes.

Oncogene	Cell signaling pathway	Tumors
abl	Signal transducer at membrane tyrosine kinase	Chronic myeloid leukemia
erbB	Signal transducer at membrane tyrosine kinase	Breast
K-ras	Intracytoplasmic protein guanosinetriphosphate (GTP) binding	Pancreas, lung
myc	Nuclear transcription factor	Acute myeloid leukemia
sis	Platelet-derived growth factor (PDGF)	?

Table 7.2. Characteristics of oncogenes and tumor suppressor genes.

Oncogene	Tumor suppressor gene
Exists in a viral form	No viral counterpart
Acts as a dominant trait, single allele mutation in tumors	Acts as a recessive trait, two alleles mutated in tumors
Gene protein promotes cell growth	Gene protein inhibits cell growth
Mutated to hyperfunction in tumors	Mutated to inactivation in tumors
Not inherited as a germline mutation; acquired as a somatic mutation	Inherited as a trait or acquired as a somatic mutation

drug therapy (as we will discuss in Chapter 8). We can recognize the DNA coding motif of a tyrosine kinase, searching the entire human genome data base for occurrences of this motif. I expect that shortly all tyrosine kinase–like receptors will have been discovered.

MYC

MYC is another oncogene, more virulent than *ABL*. Whereas *ABL* is at the beginning of the cell signaling pathway, a receptor on the surface, *MYC* is at the end. *MYC* codes for a protein that binds directly to DNA in the nucleus, affecting gene expression. *MYC* protein is equivalent to a transcription factor, as discussed in Chapter 2. Acquired mutations in *MYC* are an important step in clonal evolution of a cancer. *MYC* is mutated by a number of mechanisms including translocation, as we will see in the example of Burkitt's lymphoma. *MYC* can also be mutated by a change in its promoter control sequences. Overexpression of *MYC*, by whatever mechanism, leads to a very direct and strong signal, stimulating cell division.

Table 7.3. Examples of tumor suppressor genes.

Gene	Hereditary syndrome	Tumors
APC	Familial adenomatous polyposis	Polyps and colon cancer
BRCA1, 2	Familial breast cancer	Breast cancer
hMSH2	Hereditary nonpolyposis coli	Colon cancer
p53	Li-Fraumeni	50% of all cancers
RB1	Retinoblastoma	Osteosarcomas

Tumor Suppressor Genes

Tumor suppressor genes in general work to repair DNA damage. We certainly have not discovered all of them and only partially understand the function of some. Whereas oncogenes were discovered in association with tumor causing viruses, tumor suppressor genes were discovered in families with a hereditary predisposition to cancer. Table 7.3 lists a few examples. Tumor suppressor genes (at one brief time also called antioncogenes) functionally have an almost opposite effect from oncogenes. Their lack of function is associated with tumors, whereas oncogenes hyperfunction in tumors. These and other differences are summarized in Table 7.2. The best way to understand tumor suppressor genes is to consider some examples in some detail.

p53

The importance of the *p53* tumor suppressor gene has attracted great attention. *Newsweek* magazine named it the Molecule of the Year in 1996. The function of *p53* is to oversee the cell's repair of damage to its DNA. As we discussed in Chapter 2, *p53* holds the cell back, allowing it time to repair DNA damage at the G1/S cell cycle check point. If the damage cannot be repaired, *p53* instructs the cell to undergo apoptosis. Acquired mutations in the *p53* gene are found in 50% of human cancers. Cancer cells, once free of *p53* oversight, acquire more DNA damage and progress toward more malignant behavior. Figure 7.3 demonstrates how a cell with normal and mutated *p53* function responds to DNA damage. The Li-Fraumeni syndrome is a rare hereditary condition in which partial damage to one *p53* gene (remember there are two in every cell) results in a wide range of rare tumors. There are no known conditions with hereditary loss of both alleles of *p53* or any other tumor suppressor gene. Presumably, that much loss of DNA repair capability is not compatible with life. Acquired damage to *p53* is much more common than inherited defects. Acquired damage to *p53* results in many, many different mutations.

The function of *p53* is measured in a number of clinical cancers. Bladder and breast cancers frequently have *p53* mutations. The measurement of *p53* in those and other tumors has both prognostic value

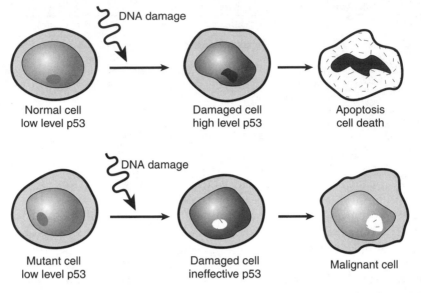

Figure 7.3. A schematic model shows alternate pathways following DNA damage in a cell with a normal *p53* gene and a cell with a mutant *p53* gene.

and, as we will see at the end of this chapter, implications for therapy. The simplest way (at the moment) to measure *p53* function is to look for overexpression of the protein by performing immunohistochemistry with an antibody to *p53* on a tissue biopsy. Most *p53* mutations, even as they disrupt function, lead to a stabilization of the normally rapidly degraded *p53* protein. Tumor cells without *p53* function stain brightly with *p53* antibody as demonstrated in Fig. 7.4. This is a breast biopsy showing damaged *p53* function in a case of invasive ductal carcinoma. We can perhaps attack these *p53* defective tumor cells, by taking advantage of exactly the defect that makes them cancer.

BRCA1

I used *BRCA1* as an example of a fairly complex gene with 23 exons in Chapter 1 as we were getting used to gene maps (Fig. 1.3). There is still a lot that we do not know about this gene, most importantly its normal function. Like many other tumor suppressor genes, *BRCA1* was discovered as a mutation in families with an inherited great risk of cancer, almost exclusively breast or ovarian cancer. One specific mutation, del185AG, a deletion of bases AG at the 185th and 186th positions, was the first discovered in this gene. This mutation is seen in about 0.9% of

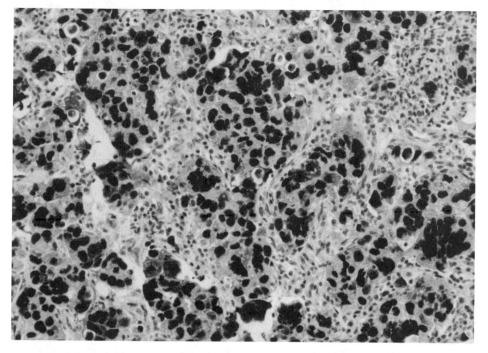

Figure 7.4. A photomicrograph of a biopsy showing invasive carcinoma of the breast stained with an antibody to the p53 protein. The cancer cells with mutated *p53* function show an accumulation of ineffective protein, indicated by black staining over the nucleus. (Courtesy of Angela Bartley, Wake Forest University School of Medicine.)

Ashkenazi Jews. The del185AG mutation results in a frame shift that stops translation of the messenger RNA (mRNA) to protein, as we saw in Chapter 2 (Fig. 2.3). There are over 200 different additional mutations in *BRCA1* that also interfere with gene function and result in a high risk of breast cancer. We can perform a clinical test for *BRCA1* mutations by sequencing the gene. With DNA on a chip technology (recall Chapter 3) this is feasible.

The clinical use of *BRCA1* testing is controversial. Cost is only one issue. Breast cancer due to hereditary predisposition makes up only 5% to 15% of all breast cancers. *BRCA1* is only one gene of several already known (including *BRCA2*) that are associated with hereditary risk. How many genes should we test? Furthermore, if a woman tests negative for *BRCA1*, she still has a standard lifetime risk of breast cancer of about 10%. A negative test should not be allowed to decrease normal clinical surveillance including self-examination and mammograms. A woman who tests positive for *BRCA1* has a very high probability (50% to 60%) of developing breast cancer in her 40s. What is the appropriate medical followup of a positive genetic test for *BRCA1*?

This is a subject of continuing controversy and an example of the complexity of genetic testing. Here, the technology is not a problem. We can sequence the gene and spot any mutation. What should we do, is a much more difficult issue. Possible therapies range from prophylactic bilateral simple mastectomies and oophorectomies to close surveillance with early and frequent mammograms.

Angiogenesis

As a neoplastic clone of cells grows, it eventually reaches a size where nutrients cannot be obtained by simple diffusion. The tumor needs a blood supply. In a cell culture model where no blood supply exists, tumor spherules will grow to about 0.5 mm in maximum size, limited by diffusion of nutrients into the center. As the spherule of tumor cells tries to get bigger, new cell growth on the surface is balanced by cell death from starvation in the center. If cancers could grow only to the size of a pinhead, they would be of no clinical significance.

The process whereby a growing tumor recruits a new blood supply from the body is called **angiogenesis**. Some would say that the clinical disease we call cancer does not start until this stage in tumor development is reached. There is much debate about to what degree the tumor is responsible for recruiting growth of capillaries. The natural repair processes of the body supplies new capillary growth in areas of tissue damage and inflammation. Some factors secreted by tumors act as growth factors on the surrounding normal tissue, invoking both the proliferation of fibroblasts and capillaries. If tumors had a specific angiogenic growth factor, then inhibition of this factor could be an important anticancer therapy. Many attempts at blocking tumor angiogenesis have resulted in only limited success, but the promise is still very great.

What is very clear is the realization that a growing tumor must overcome tissue barriers. Figure 7.5 shows a small squamous cell carcinoma of the bronchus of the lung just beginning to invade deeper tissues. This small cancer has begun to recruit budding capillaries. The endothelial cells are proliferating in response to local growth factors. An intense angiogenic stimulus is present. We can envision that this 1cm diameter tumor will grow at a rate that is primarily determined by its blood supply, rather than by its intrinsic mitotic rate. New therapies for lung cancer include drugs such as thalidomide, which may inhibit angiogenesis in a nonspecific fashion.

Tissue Invasion and Metastasis

In addition to developing a blood supply, tumors must overcome tissue barriers in order to invade. The hallmark of cancer as recognized by the pathologist under the microscope is invasion through local tissue bar-

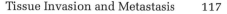

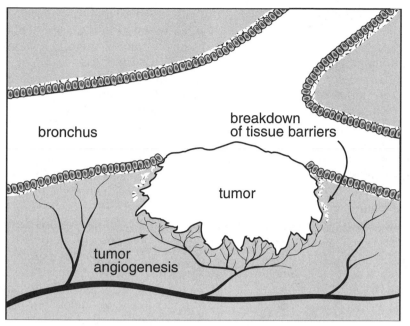

Figure 7.5. A tumor develops beyond microscopic size by the processes of angiogenesis and invasion through tissue barriers.

riers.[2] Most cancers have a recognizable **carcinoma in situ** stage where the tumor is confined to the epithelial layer. Carcinoma in situ is curable by surgical excision. The goal of early clinical detection of cancer is to find tumors at this stage. Unfortunately, cancers are usually detected only at more advanced stages. About one-third of cancer patients present as advanced-stage disease with distant metastases. Another one-third present with locally invasive disease, but no clinical evidence of metastasis. Clinical experience has taught us that for many forms of cancer such as lung or pancreas, there is a high probability that undetected micrometastases are already present when the tumor is locally invasive. We are left with only one-third of patients who are diagnosed with truly localized disease that is surgically resectable.

The process of tissue invasion and metastasis is therefore very relevant to the clinical management of cancer. For most cancers derived from epithelium (such as colon, cervix, or breast) we can identify patients at the stage of carcinoma in situ. At this point, the clonal neo-

---

[2]This is cellular medicine, a concept that has reigned for several centuries, beginning with the realization that the body is made up of cells organized into tissues and organs. Many diseases including cancer were found to be a dysfunction at the cellular level. We are now at the dawn of molecular medicine, with the deeper insight that cellular dysfunction results from errors in information as encoded in DNA.

plasm has not extended beyond the retaining basement membrane that underlies all epithelium. The basement membrane is an extracellular structure made up of collagens and glycoproteins including laminin and fibronectin. Epithelial cells have receptors for these molecules. The basement membrane provides a barrier and a point of anchorage for the epithelial cells. Normal cells cannot grow and function without this point of attachment. Cancer cells progressively lose this dependency on the basement membrane. To invade, the tumor must breach the basement membrane, usually by secreting collagenases and other enzymes that destroy connective tissue elements. These enzymes include plasminogen, cathepsins, and hyaluronidase. The invasion of the connective tissue by the tumor is a subversion of the body's normal inflammatory and repair process.

At the next stage beyond carcinoma in situ, the tumor extends through the basement membrane. The enzymes that break down the local tissue barriers may serve as biomarkers of the tumor. Cathepsin D is a proteinase that is a marker correlated with increased biological aggressiveness in breast cancer. Other biomarkers are related to surface molecules on cancer cells that correlate with their loss of dependence on attachment to the basement membrane. Like angiogenesis, tumor invasion through tissue barriers is a possible point at which to target new anticancer therapies. Some of the reported effects of drugs such as heparin and antiinflammatory agents on cancer may work at this stage.

As the tumor cells invade connective tissue past the basement membrane, they gain the potential to migrate further into lymphatic or venous channels. This is the start of metastasis, spread to distant parts of the body. To establish a metastasis, however, the tumor cells must survive the body's immune system. They must also find a tissue environment that will support their growth needs. Some tumors predictably metastasize to the most available capillary bed. Breast cancer goes to lung; lung cancer goes to brain. Other tumor cells grow best in more specific environments. The bone marrow supplies a good microenvironment for metastases. The characteristic special patterns of metastasis, such as stomach cancer spreading to the ovaries, must indicate some special hormonal or biochemical support of the tumor cells.

Colon Cancer: Model of a Multistep Process

Colon cancer has a well-defined, multistep evolution defined both by visible and by molecular events as demonstrated in Fig. 7.6. Colon cancer starts as a polyp that enlarges and becomes more dysplastic. Carcinoma in situ within a polyp precedes early invasive cancer. A very rough estimate is that polyp to cancer takes 5 years. The molecular lesions of colon cancer appear years before invasive cancer. The cells lin-

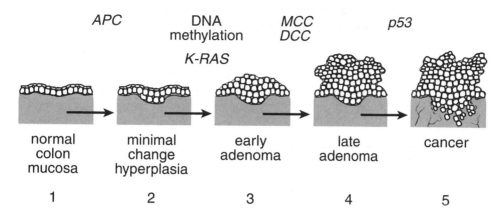

Figure 7.6. The molecular evolution of colon cancer is a series of mutations that parallels visible changes.

ing the colon are bathed in a lifelong effluent of bacteria and break-down products of digestion that produce carcinogens.

In Fig. 7.6, I have shown the molecular evolution beginning with a mutation in the *APC* tumor suppressor gene. At this early stage, we may see no visible change in the colonic epithelium. The next step is a change in the methylation pattern of DNA within these cells. Methyl groups can be attached to DNA like pasted notes on a memo. They are in part a record of the history of that piece of DNA. DNA that has been "read" tends to be more methylated. Another event likely at this early stage is a point mutation in *K-RAS*. By now, I would expect the neo-plastic process to be visible as an early adenoma when viewed through the endoscope. Over time, the next step is a larger villous ade-noma. In parallel with this development are changes in the *DCC* (deleted in colon cancer) and *MCC* (missing in colon cancer) genes. If untreated, the late-stage villous adenoma has a high probability of progressing to an invasive colon cancer. Invasive colon cancer almost always has a deletion mutation in *p53*. The cancer is now capable of invading surrounding stroma presumably due to a loss of cell surface receptors. These surface receptors serve as tissue anchors for normal colon cells. Without *p53* function the tumor cells rapidly acquire fur-ther mutations. The tumor now elicits angiogenesis to increase its blood supply through a perversion of the normal tissue repair process.

The molecular scenario that I have just described is a common path-way in the multistep development of colon cancer. Other gene muta-tions can occur that will lead to the same result. There are not always five molecular steps. Other combinations of multiple molecular events offer different paths to the same biological outcome.

Squamous Cell Carcinoma of the
Uterine Cervix: Model of a Viral Role in Cancer

Squamous cell carcinoma of the uterine cervix is a disease that can be detected by Pap smears, and when detected early, cured with cone excision of the cervix. We now know that this cancer is almost entirely caused by persistent infection of the cervix with human papilloma virus (HPV). As we will see, this immediately suggests even better screening by viral detection, instead of a Pap smear. More importantly, we can hope to cure the HPV infection with a DNA vaccine.

Cervical cancer takes years, even decades, to progress from a slightly abnormal Pap smear to invasive cancer. Figure 7.7 demonstrates the time course of abnormal Pap smears progressing to cancer. The Bethesda nomenclature of ASCUS (atypical squamous cells of uncertain significance), LGSIL (low-grade squamous intraepithelial lesion), HGSIL (high-grade SIL), and CIS (carcinoma in situ) is used. The multistep pathway of cancer development begins with infection of the squamous epithelium of the cervix by HPV and is reflected on a Pap smear as ASCUS. In these first stages of evolution of cervical cancer, there is a high probability that the abnormality monitored by Pap smears will return to normal. In a few percent of cases, the epithelium of the cervix will become severely dysplastic (CIS), and if untreated progress to invasive carcinoma. About 15% to 25% of women will at some time in their life have detectable HPV on a cervical swab, the incidence peaking between 15 and 30 years of age. Most of the infections will resolve as the body develops immunity to HPV. The associated Pap abnormalities will also resolve. The longer the HPV infection persists, the greater the likelihood of progression to cancer. We know something about how this comes about.

HPV

The human papilloma virus (HPV) contains circular, double-stranded DNA, 7,900 base pair (bp) in length contained within a capsid. There are at least 70 strains of HPV with widely varying disease-causing potential (Table 7.4). HPV infects epithelial cells by direct contact and is thus predominantly a sexually transmitted disease in the urogenital areas. HPV infection occasionally causes warts (condyloma) but more commonly results in an asymptomatic infection. Urogenital warts are usually flat and not easily recognizable, especially on the penis.

Much of the viral genome of HPV contains the usual genes necessary to maintain a virus. The HPV virus, like many others, borrows much of the genome from the cell it infects in order to make all of the enzymes it needs for replication. Two very special genes of interest to us, *E6* and *E7,* have transforming or oncogenic potential for human cells. The E6

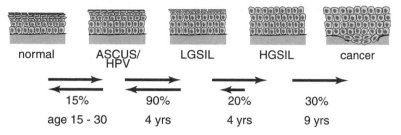

Figure 7.7. A diagram shows the progression of cervical dysplasia as seen on Pap smears. The time spans and probabilities for passing between stages are indicated at each step.

and E7 proteins can inactivate the tumor suppressor genes *p53* and *RB1*. When *p53* is inhibited by interference from the E6 protein, the infected squamous epithelial cell can enter the S phase of the cell division cycle without being checked. This greatly increases the possibility that errors in DNA synthesis will occur resulting in mutations. Similarly E7 protein inhibits the *RB1* tumor suppressor gene. Those strains of HPV virus with the greatest oncogenic potential are exactly the strains that most completely interfere with *p53* and *RB1* function.

Why would HPV have genes that interfere with our tumor suppressor genes? The virus needs to block these genes in order to live within the epithelial cells. The virus represents abnormal DNA, which may alert the *p53* system into blocking cell division. The virus needs the cell division cycle in order to propagate! So it has evolved proteins E6 and E7 to gain control. Exposure of the squamous epithelium to HPV results in infection of some of the basal cells. We can detect this infection by in-situ hybridization with molecular probes for HPV. This is demonstrated in Fig. 7.8, a photomicrograph of a cervical biopsy. This biopsy was graded as mild squamous dysplasia and was associated with a LGSIL Pap smear. If this patient develops immunity to HPV before a critical amount of damage occurs to the DNA in the infected cells, then the virus infection will terminate and the dysplasia will resolve. If a more oncogenic strain of the virus persists for a longer time, cancer will result.

Anti-HPV, a Possible Cancer Vaccine

The immunity necessary to provide protection for some viruses is not easily obtained. There are many viral and host factors that we simply do not understand. Most traditional vaccines are derived from a piece of the viral coat and invoke humoral immunity derived from B cells. A new class of vaccines based on DNA is being studied for intracellular viruses with a long latent phase of infection. DNA vaccines have the

Table 7.4. Human papilloma virus (HPV) strains and cervical carcinoma risk.

HPV strain	Risk	Viral protein interaction with tumor suppression gene	
6,11	Low	E6 does not bind *p53*	E7 weakly binds *RB1*
31,33,35,51,52	Intermediate		
16,18,45,56	High	E6 binds *p53*	E7 strongly binds *RB1*

potential of invoking cellular immunity. DNA vaccines require a vector to carry the DNA into the cell. These vaccines may use live virus vectors or other means to carry a piece of DNA across the cell membrane. Once inside the cell, the engineered pieces of DNA in the vaccine code for viral proteins. When these proteins are produced inside the cell, they invoke T-lymphocyte–mediated cellular immunity.

HPV (like HIV) is a virus that can be overcome by cellular immunity. While we commonly think of a vaccine as useful only in preventing disease, HPV is a chronic infection and there is plenty of time. We can give a DNA vaccine against HPV after the infection is detected. Several anti-HPV vaccines are in various stages of development and clinical trials. The questions to be answered by the clinical trials are, Does the vaccine reverse the infection? and Does it prevent or reverse the development of dysplasia and thus prevent cancer?

Lymphoma and Leukemia

In Chapter 6 we learned about the ingenious and somewhat complicated way that lymphocytes generate a nearly infinite array of antibody diversity. Gene rearrangement through multiple splicings of the members of the immunoglobulin gene family works, but with errors. Lymphoma results when the error involves an oncogene! In fact, insofar as we know, all leukemias and lymphomas are due to erroneous cutting and splicing of chromosomes with insertion of an oncogene. Let's consider three specific examples.

Chronic Myeloid Leukemia

Chronic myeloid leukemia (CML) was the first malignancy associated with a specific clonal marker, the t(9;22) Philadelphia chromosome, as described in the 1960s. At the start of the molecular era in the 1980s, researchers in Rotterdam asked, "When chromosomes break in tumors, do the breaks always happen at the same place on the DNA molecule, or are the breaks widely distributed?" They chose to study CML. They demonstrated that the break on chromosome 22 was always within a

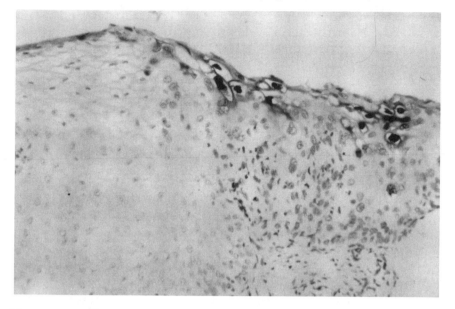

Figure 7.8. A photomicrograph of an in-situ molecular probe for HPV strain 18 demonstrates intracellular virus in this cervical biopsy. The dark-staining nuclei in the vacuolated cells of the upper epidermis contain virus.

rather small area on the chromosome that they named the break-point cluster region (bcr). We now know that this region is within a gene that is named *BCR*. The molecular biology of CML shows how a translocated oncogene results in a tumor. To understand the mechanism, let's begin with an analogy that I believe is useful.

Registrar's Mutation

Consider a medical school library (as we did in Chapter 1 for a model of the information content of the human genome). There are many books, only a few of which deal with the organizational structure of the medical school. In one of these, *Instructions for Enrollment, Course Registration, and Graduation*, there is a mutation. Near the end of the book, the text has been erroneously modified by a translocation of a paragraph from near the beginning. In the original correct version, the fourth year medical student is instructed to take his or her grade transcript to the registrar in exchange for a diploma. In the mutated version, immediately following "take your transcript to the registrar," the words "and receive your first-year course assignments" have been inserted. The correct instruction "and exchange for a diploma" has been overwritten by a translocation from the instructions for entering students. What is the result of the registrar's mutation? Fourth-year students, instead of leaving and progressing to their mature function as physicians, return to the developing function of first-year students. Within a few years, the classrooms are overcrowded inside the medical school, while outside there is a shortage of physicians. Let's compare this analogy to the molecular and clinical features of CML.

BCR/ABL

The molecular event of the t(9;22) translocation is the fusion of the *ABL* oncogene from chromosome 9 into the *BCR* gene on chromosome 22 (as shown in Fig. 7.9). The *ABL* oncogene has two alternate first exons, 1a and 1b. In the normal expression of *ABL,* only one of these alternate exons is spliced to the downstream exons 2 through 11. There is a splice acceptor site in front of exon 2, and therein seems to lie the problem. This splice acceptor site is "promiscuous" in that it accepts donors other than the expected upstream exon 1a or 1b. For reasons that still elude us, a portion of *BCR* can be matched to this site. The chromosomal translocation that results from this faulty joining produces the *BCR/ABL* fusion gene.

   ABL is an oncogene that produces a tyrosine kinase that is bound just under the cell membrane as part of a growth factor receptor. The *BCR/ABL* fusion gene produces a hyperactive form of the tyrosine kinase that makes the receptor overly sensitive. CML results from this mutated *BCR/ABL* oncogene and the loss of normal function in the cell signal pathway. In Chapter 8, I will describe an engineered molecule, the drug STI571 (Gleevec Novartis, Basel, Switzerland), that specifically blocks the abnormal enzyme function of *BCR/ABL*. This drug was released for the treatment of CML in the summer of 2001.

   The loss of normal response by the growth factor receptor results in something like the registrar's mutation. Bone marrow cells stop maturing into normally functioning granulocytes. The marrow fills up with myeloid cells that fail to differentiate fully. All leukemias seem to follow this pattern; failure of maturation with clonal expansion and packing of the bone marrow with simultaneous loss of the production of normal mature-functioning cells. The chromosomal translocations and the genes involved are different for other forms of leukemia, but the theme remains the same.

Follicular Lymphoma

Follicular lymphoma offers another example of a translocated oncogene, this time leading to a low-grade and indolent neoplasm. The t(14;18) translocation, commonly seen in this form of lymphoma, results in the incorrect insertion of a gene called *BCL-2* from chromosome 18 into the immunoglobulin heavy chain gene, *IgH,* on chromosome 14. The breakpoint where *BCL-2* inserts varies from patient to patient, resulting in sev-

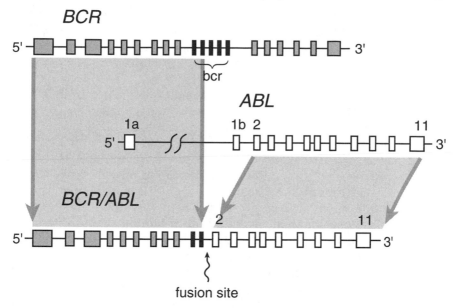

Figure 7.9. A gene map of *BCR/ABL* fusion shows the *ABL* exons spliced to the tail of the *BCR* gene.

eral different mRNAs from this abnormal fusion gene. This may explain some of the biological variability in lymphoma.

Whenever a lymphocyte with the *IgH/BCL-2* fusion tries to make immunoglobulin, it produces a bcl-2 protein. The bcl-2 protein inhibits apoptosis! The abnormal lymphocyte does not undergo programmed cell death. Although the tumor cells have only a low proliferating fraction, they hang around nearly forever, leading to a slow expansion of the tumor. This is exactly what we observe clinically.

Burkitt's Lymphoma

Burkitt's lymphoma is a high-grade malignancy that results from a different oncogene, *c-MYC*, inserting into the immunoglobulin *IgH* gene, due to a t(8:14) chromosomal translocation. The process is like that in follicular lymphoma, except that *MYC* is much "tougher" than *BCL-2*. *BCL-2* inhibits apoptosis and cell death. *MYC* is a DNA transcription factor that promotes cell division. Burkitt's lymphoma can also result from other splices of *MYC* inserting into either the κ or λ immunoglobulin light chain genes on chromosomes 2 or 8. Anytime *MYC* jumps from its home on chromosome 8 to an immunoglobulin gene, this high-grade malignancy results. When the abnormal lymphocyte tries to make an immunoglobulin, it instead produces c-myc protein that binds to DNA to promote division. The tumor grows rapidly with strong autocrine stimulation. The insertion of *MYC* into an immunoglobulin gene in Burkitt's lymphoma explains only part of the pathogenesis of this disease. The rest of the story is more complex and involves the immune system. Again we will see that cancer is a multistep process with multiple etiologies.

Burkitt's lymphoma occurs at a much higher frequency in immunosuppressed patients with T-cell deficiency. This is notable in HIV and malaria. Sir Dennis Burkitt described the lymphoma named after him in areas of Africa that were known for a high incidence of malaria. He could have asserted that mosquitoes "caused" lymphoma. However, Burkitt, being a good epidemiologist, emphasized only the association. We know that mosquitoes transmit the microorganism that causes malaria. Malaria in turn suppresses T-helper lymphocyte immune function. T-helper lymphocytes help hold B-lymphocyte proliferation in check. When T-lymphocyte function fails, the incidence of Burkitt's lymphoma (and other malignancies) soars.

This appreciation of the role of the immune system in holding B-cell clones in check was not understood until the era of AIDS. HIV infection, like malaria, suppresses T-helper lymphocyte function. When this was discovered, physicians knowledgeable about malaria correctly predicted that HIV patients would also demonstrate a high incidence of Burkitt's lymphoma.

But there is yet another important step in the pathogenesis of Burkitt's lymphoma, infection with Epstein-Barr virus (EBV). In Africa, virtually all infants have experienced EBV infection prior to 1 year of age. EBV causes a polyclonal expansion of B lymphocytes. As immunity develops, this expansion is suppressed by T lymphocytes. When T immunity fails secondary to malaria or HIV, the EBV-infected B lymphocytes again proliferate. This late proliferation of EBV-infected lymphocytes is prone to mistakes in gene rearrangement. This leads to a higher probability of a translocated *c-MYC* gene.

Knowing all these steps, what can be said to "cause" the tumor? Figure 7.10 summarizes the multiple steps leading to Burkitt's lymphoma. The first causal event is EBV infection, which occurs in infancy. The next causal event is the malarial mosquito that leads to an infection that suppresses T-lymphocyte function. This is followed by a chromosomal translocation that activates *MYC*. So when a patient asks, "What caused my lymphoma," what can we answer?

Molecular Therapy of Cancer

Cancer results from a series of mutations in oncogenes and tumor suppressor genes that disrupt controls on cell growth. This knowledge provides a rational basis for the development of entirely new avenues of anticancer therapy. The most direct approach, but also the most difficult,

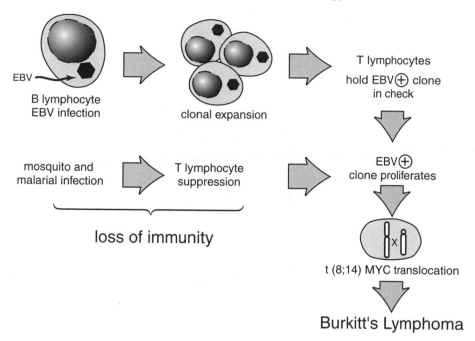

Figure 7.10. Multiple steps in the evolution of Burkitt's lymphoma involve Epstein-Barr virus (EBV) infection, loss of T-cell immunity, and a chromosomal translocation of *MYC*.

is to fix the mutations in DNA through genetic engineering. If we cannot do that, we can try to block abnormal gene function by interfering with transcription or translation. Alternatively, we can intervene at a later step with antibodies to abnormal oncoproteins. Figure 7.11 summarizes these possible approaches. Let's consider a few specific examples.

Antisense Oligonucleotide Therapy of *c-MYC*

Antisense oligonucleotides (recall Chapter 3) are a new class of drug that block specific gene expression. Consider a case of Burkitt's lymphoma as we just discussed with an activated *c-MYC* gene due to a chromosomal translocation and fusion with the immunoglobulin heavy chain gene. We can synthesize a small piece of DNA complementary to the sequences of the fusion region including the start of *c-MYC*. We will make our piece complementary to the transcribed mRNA sequences, some of which I listed in Fig. 1.5, anatomy of the *c-MYC* gene. This piece is antisense DNA in that it binds to the sense mRNA sequences. The binding to mRNA stops the expression of this abnormal fusion gene and the stimulus to the tumor cells to divide. We can either give the drug repeatedly or, in theory, we can make that block permanent by transfecting cells with the antisense sequence. Antisense offers a very specific therapy based on our

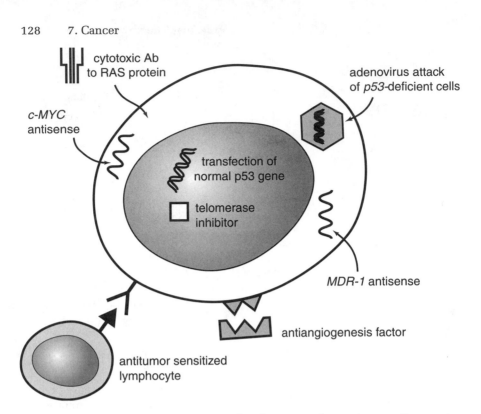

Figure 7.11. There are many new molecular targets for anticancer therapy to correct, block, or destroy abnormal function.

exact knowledge of the genes at fault in cancer. The technical problems at the moment are in getting the molecule inside the cell and preventing its too rapid breakdown.

Adenovirus Killing of *p53*-Deficient Cancer Cells

Half of all cancers have a mutation in the *p53* tumor suppressor gene. Can we turn the tables and take advantage of this defect in DNA repair that allows cancers to progress as a target for therapy? Adenovirus, the infectious agent that causes many minor upper respiratory infections (colds), has genetic machinery that allows it to subvert *p53* function in human cells. The adenovirus has a gene that allows it to override *p53* function. This aids its ability to replicate in human cells. A genetically engineered adenovirus that does not have this gene has been created as an anticancer agent. This strain can propagate only in *p53* deficient cancer cells! Cancer cells are the targets for infection by this strain of adenovirus, where normal cells are spared. The cancer cells burst when infected and die. Soon, however, the body develops immunity to the adenovirus and eliminates the infection. Clinical trials with this genetically engineered adenovirus are under way in cancers of the pancreas, mouth, and other sites. The idea to use a biological vector to kill selectively, based on the damaged genome present in cancer cells, is a powerful example of molecular based therapy.

Antibodies to Oncoproteins and Tumor Antigens

The *RAS, p53,* and *HER2/neu* oncoproteins are targets for antibody therapy because of their abnormal overexpression in cancer cells. A commercially available antibody to *HER2/neu* is used as a treatment of metastatic breast cancer when this gene is overexpressed. Antibodies to *p53* and *RAS* are in various phases of clinical trial. An extension of this concept is to make more specific antitumor antibody vaccines of both the humoral and cellular type from a patient's tumor. The cellular immunity offered by killer T lymphocytes is one of the more promising areas. The lymphocytes are "trained" outside the body to become targeted to the tumor, and then reinfused along with cytokines that promote their function. Therapy based on manipulation of the immune system suggests many, many new approaches now that we have the ability to synthesize cytokines and other immune modulators.

Summary

We have discovered that cancer is essentially a molecular disease caused by acquired errors in DNA. Mutations accumulate in multiple steps in oncogenes and tumor suppressor genes leading to unregulated cell growth. The process usually takes years to develop, as we saw in the examples of colon and cervical cancer. New therapies directed against cancer span a range of modalities. We can stop infection of the cervix by HPV with a DNA vaccine. We can block the abnormal fusion genes that result from chromosomal translocations in leukemia and lymphoma with tailored drugs. Finally, we can hope to manipulate the immune system to recognize cancer cells as foreign. Major advances in the early detection and in the treatment of cancer will follow our new molecular understanding of this disease.

Bibliography

Bartley, AN, Ross DW. Validation of p53 immunohistochemistry as a prognostic factor in breast cancer in clinical practice. *Arch Pathol Lab Med* 2002 (in press).

Bishop JM, Weinberg RA, eds. *Molecular Oncology.* New York: Scientific American, 1996.

Cancer Genome Anatomy Project (CGAP). *http://www.ncbi.nlm.nih.gov/ncic-gap.*

Erlich M. *DNA Alterations in Cancer.* Natick, MA: Eaton, 2000.

Lowry DR, Wolff L. Molecular aspects of oncogenesis. In: Stamatoyannopoulos G, Majerus PW, Perlmutter RM, Varmus H, eds. *The Molecular Basis of Blood Diseases*, 3rd ed. Philadelphia: WB Saunders 2001:791–831.

Pulverer B, ed. Insight cancer. *Nature* 2001;411:355–395.

Ries LAG et al. *SEER Cancer Statistics Review, 1973–1993: Tables and Graphs.* Bethesda: National Cancer Institute, 1996.

Ross DW. *Introduction to Oncogenes and Molecular Cancer Medicine.* New York: Springer-Verlag, 1998.

8   Molecular Therapeutics

Overview

Molecular therapies have been presented throughout this book. In this chapter, I will demonstrate how molecular technology is revolutionizing the way in which we discover drugs. Many companies have invested their futures in harvesting drugs from the human genome. Pharmacogenetics is one aspect of this revolution. Pharmacogenetics specifically looks at how genetic differences influence the response of a patient to a drug. Rather than consider an individual's response to a drug as idiosyncratic, we can look at it as a potential advantage. The single nucleotide polymorphism (SNP) map of each patient will influence which drug to prescribe for a disease in that person. We will begin our use of SNP mapping with the cytochrome enzyme system of liver cells, as this is the most important modifier of drug action that we have yet discovered.

I will conclude this chapter with other examples of new molecular therapies. While molecular medicine is too young to predict how it will grow, we can envision some of its potential. As examples of the possible future, I will look at some new therapies in development: gene transplantation, tailored drugs created by computer modeling, drugs raised in plants, biomaterials, and the transplantation of embryonic stem cell derived tissues. This chapter provides a sense of what will be possible in molecular therapeutics.

Pharmacogenetics

Pharmacogenetics is the study of how genetic differences influence the variability in patients' responses to drugs. The hope is that the genetic uniqueness of individuals can be used to predict how they will respond to certain drugs. A database of genetic polymorphisms correlated to clinical outcome may be one means by which the objective of pharmacogenetics is achieved. We begin by looking at components of the problem.

SNPs

Single nucleotide polymorphisms (SNPs, pronounced "snips") are single base differences in the DNA of individuals. Typically a polymorphism at a SNP site occurs with a frequency of 1% or greater. If the frequency is lower than 1% or if the alteration is known to be the cause of a disease, then we call the change a mutation rather than a SNP. For example, hemoglobin S (due to a single base change) occurs with a frequency greater than 1%, but it is the cause of a disease and is therefore considered a mutation. In Chapter 1, we noted that the Human Genome Project has mapped about 2.5 million SNPs, or about 1 per 1,000 base pair (bp). SNPs are markers of genetic differences but not the cause of the difference.

SNPs are very useful in the field of pharmacogenetics. We can easily perform a SNP genotype of a particular chromosome region using DNA on a chip or other high-speed sequencing technology. This "SNP-shot" gives us a picture of the individual's genetic makeup. Figure 8.1 shows how we can use a SNP-shot to determine which patients are likely to respond to a given drug therapy for a specific disease or syndrome. We match up SNP-shots with clinical outcome. In some instances a specific SNP genotype found in a set of individuals will be highly correlated with a good response to therapy. The converse is also true. We sometimes find that a certain SNP-shot correlates with an adverse drug reaction. Once we have this knowledge, correlating SNP genotype with outcome, we can then use SNP mapping as a predictor prior to giving a drug.

A SNP genotype, although new, is not entirely different from what many physicians already do. Physicians use the experience of outcome in previous patients to estimate whether the current patient is likely to benefit from a specific treatment. Sometime this prior experience is genetic. If you believe that red-headed people are more likely to be allergic to penicillin, you will use this supposition to modify your prescribing of antibiotics. A SNP genotype greatly extends and adds precision to our genetic assessment of an individual. Many people working in pharmacogenetics strongly believe that when SNP technology is widely utilized, we will greatly increase the efficacy of certain drugs and greatly decrease the incidence of adverse drug reactions.

Cytochrome P-450 System

The cytochrome P-450 (CYP) family of heme-containing enzymes oxidizes many lipophilic chemicals. CYP is responsible for the metabolism of up to 50% of drugs. CYP3A4, which is the major microsomal CYP in the liver, is an example of one of the multiple enzymes in this

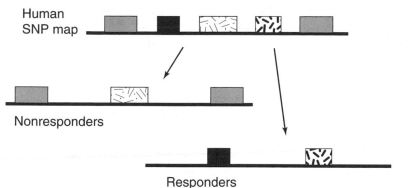

Figure 8.1. Single nucleotide polymorphism (SNP) mapping determines who will respond to drug treatment (see also Roses, 2000).

group. CYP3A4 is a modulator of most of the endogenous steroid hormones. CYP3A4 also breaks down half of all drugs, making them water soluble for excretion by the kidney. The activity of the CYPs is itself modulated by the amount of substrate present, as well as many physiological factors including physical activity, stress, and inflammation. In pregenomic studies, CYP activity in individuals was sometimes probed with a test substance like caffeine to estimate response to a drug.

All of the CYP enzymes are highly polymorphic (i.e., there are a lot of SNPs) as well as existing in multiple alleles. It is therefore no surprise that CYP3A4 is a major target of pharmacogenetics. This is a central point in drug metabolism. For example, many of the cardiac statins (lovastatin, simvastatin, atorvastatin) function via CYP activity. This means that other modulators of CYP will create drug–drug interactions with the statins. These drug–drug interactions were previously described as "idiosyncratic," meaning unpredictable. There is a belief that if we understand an individual's CYP genotype or SNP profile, these drug interactions will no longer be unpredictable. In fact, we will be able to tailor individual therapy to a much higher degree of predictable response. We can imagine a patient having a copy of his CYP genotype as part of the medical record, even perhaps encoded on a pharmacy card. A database of genotypes will predict the correct drug, dosage, and interaction with other drugs as each prescription is filled.

Drug Development

The discovery of new drugs until very recently was a slow, even tedious search, cataloging natural substances with potential beneficial effects. Drug discovery began with folklore; the bark of the willow tree relieves fever (aspirin through the prostaglandin pathway blocks mediators of inflammation); a tea of the foxglove plant cures dropsy (digitalis improves cardiac output minimizing abdominal ascities). Serendipity also led to new drugs, such as Fleming's discovery that the growth of a mold inhibited bacteria on his Petri dishes. Modern drug development now moves into synthetic molecules designed to fit computer-simulated targets.

High-Throughput Screening Technologies

The human genome has been called a mine of valuable products waiting to be developed. The potential for discovery has been one of the issues that has created the controversy of who "owns" the genome? Whatever the outcome, the mining of the genome has begun in earnest. *Bioinformatics, protein modeling,* and *high-throughput drug screening* are buzzwords for computer programs that chew through the 3 billion bp human genome database looking for a product. We know what the protein structures of important drug targets, such as surface receptors, kinases, enzymes, and transporters, look like. We work backward and anticipate what the DNA code for these structures must be. Then we engage a search program to find all occurrences of DNA sequences that are close to our target sequences. These newfound sequences are then each examined in more detail. The entire open reading frame (ORF) around a found sequence is modeled into a possible protein, and its structure is guessed at. From the supposed structure, we further guess what drugs, existing or theoretical, might interact with our guessed-at virtual protein.

Figure 8.2, in a very schematic fashion, diagrams these processes, in the style of a flowchart. In the recent past, understanding even one protein–drug interaction at this level would have amounted to a research proposal for a postdoctoral fellow. "One molecule, one postdoc" used to be the aphorism. This phrase is in itself a parody of the original "one gene, one enzyme" pronouncement of the first molecular biologists. Now, the process that leads to drug discovery is something that might occur within minutes inside a computer. Tens of thousands of possible drug targets can be considered by a computer program under the supervision of an insightful and knowledgeable research lab. We are in the science-fiction–like position of watching on a computer screen an unknown protein binding with an unknown drug. If we like what we see, then we can begin a much longer process of turning a virtual model into real chemistry.

STI571 (Gleevec)—a Specific Inhibitor of *BCR/ABL*

In Chapter 7, we discussed the Philadelphia chromosomal translocation and its associated gene rearrangement, *BCR/ABL,* which is specific for chronic myeloid leukemia (CML). This was the first chromo-

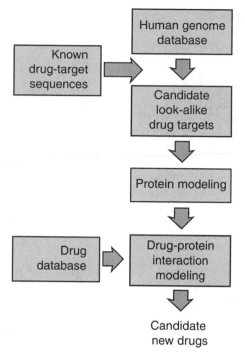

Figure 8.2. The new computer-driven process of drug discovery searches the human genome database using a knowledge of known drug targets and known drugs. Computer modeling constructs candidate look-alike interactions between yet-to-be-discovered targets with new drugs. If we like the virtual reality of the computer model, we proceed with the actual biochemical discovery.

somal translocation and later the first clonal molecular defect discovered in a tumor. It seems appropriate that one of the first specifically engineered molecular drugs would be for CML. Animal and cell culture studies have clearly shown that the *BCR/ABL* rearrangement results in the expression of an abnormal tyrosine kinase with different kinetics and substrate acceptance than a normal *ABL* tyrosine kinase. Using knowledge of the abnormal molecule, researchers sought to create a drug that would block only the abnormal kinase seen in leukemic cells. The successful result is STI571 (signal transduction inhibitor 571), also called Gleevec. It is a synthetic small molecule designed to fit in the kinase "pocket" of the BCR/ABL protein. Figure 8.3 shows the interaction of the *BCR/ABL* tyrosine kinase with its substrate. This is a two-dimensional diagram of the molecular interactions leading to phosphorylation of the substrate as it occurs in a leukemic cell. The drug molecule STI571 fits into the pocket on *BCR/ABL* where adenasine triphosphate (ATP) would normally go, thus blocking the abnormal function of the BCR/ABL fusion protein. (see also Goldman and

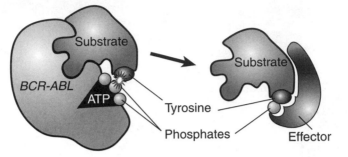

Figure 8.3. Molecular fit of STI571 (Gleevec) blocks the abnormal *BCR/ABL* kinase in leukemic cells. ATP, adenosine triphosphate.

Melo, 2001). A more accurate picture than that shown in Fig. 8.3 would be a three-dimensional picture annotated with binding attractions between the component proteins. Computer modeling of the three-dimensional form of proteins visualizes the potential target pockets for drugs like STI571. The molecular structure of STI571 is sufficiently specific to block *BCR/ABL* kinase, but has limited inhibition for the many other tyrosine kinases in our normal cells.

Initial preclinical studies completed in 1998 led to fast-track trials in patients, and the rapid Food and Drug Administration (FDA) approval of STI571 in May 2001. This drug is indicated for the treatment of the chronic phase of CML. It has also shown activity in gastrointestinal stromal tumors (GISTs) and perhaps any tumor that displays the c-kit antigen. It is hoped that STI571 is an example of how the molecular basis of a disease can be exploited to tailor-make a drug.

Gene Therapy

Gene therapy is at the cusp of what molecular medicine is all about. We are on the verge of curing illnesses that are due to inherited or acquired genetic mutations, by fixing the gene itself. In even more ambitious plans, we may transplant genes to improve physiology to a state better than the norm. It is possible that we will improve our handling of lipids and carbohydrates to ameliorate heart disease and diabetes If diet does not work, we will give our patients a pill to allow them to handle their excesses.

Gene therapy is in development on many fronts. The technical problems can be overcome, but medicine needs some experience with this entirely new form of treatment to understand the unanticipated side effects and the ethical issues. Let's consider a few examples of current gene therapy, simple compared to what the future will bring, but dramatic enough for the moment.

Miniexamples

Hemophilia and Factors VIII and IX Replacement

Hemophilia A and B (hereditary X-linked deficiency of factor VIII or IX) has traditionally been treated by repeated infusions of human plasma–derived replacement coagulation factors. This therapy, in addition to being invasive, puts the patient at risk for serious infections, especially hepatitis. Molecular technology has recently produced recombinant factors VIII and IX as drugs for replacement therapy. This removes the risk of infection inherent in using human or animal derived protein. The ideal therapy, however, would be gene transplantation. This appears to be at hand. Several clinical trials of gene therapy for hemophilia are in progress. A cloned construct of coagulation factor IX gene in an AAV vector (AAV-CMV-hF.IX, Coagulin-B, Avigen, Inc., Alameda, California) has been studied in a phase I trial at University Hospital of Philadelphia and Stanford University Hospital; three patients showed a modest clinical response. More importantly, no toxicity or inhibitory antibodies to the protein were observed. A plasmid construct of the factor VIII gene has been used to transfect cultured skin fibroblasts from patients with hemophilia A. The gene transplant occurs in a culture dish of the patient's fibroblasts in the laboratory. The plasmid is forced into the cells by electric shock. The transgenic fibroblasts (1 to 4 $\times$ 10^6 cells) are then implanted in the patient's omentum. In six patients studied, none developed inhibitors against factor VIII and four showed clinical improvement.

Immunotherapy of Lymphoma/Leukemia

Lymphoma and leukemia cells usually escape immune detection. An adenovirus vector carrying a cell surface antigen (CD40) has been used to transfect the neoplastic cells of patients in short-term cell cultures. The vector causes CD40 to be expressed on the leukemic cells. When they are returned to the patient, the transduced leukemic cells excited a cytotoxic T-cell immune response. The modified cells essentially served as an individually tailored vaccine against the patient's tumor.

Suicide Gene Therapy of Tumors

Tumor cells can be selectively killed by transfecting them with a suicide gene. The thymidine kinase gene cloned from the herpes simplex virus (*HSVtk*) has been used in several clinical trials. An example is transfecting dividing brain tumor cells with a retroviral vector construct containing *HSVtk*. After the transfection, the antibiotic ganciclovir is given.

This drug is altered by the *HSVtk* into an active lethal form that kills the tumor cell. In addition, there is a considerable "bystander effect" in that nearby tumor cells are also killed, even if they did not take up the retroviral vector containing *HSVtk*. This bystander effect is probably mediated by tight cell junctions between the infected tumor cells and their neighbors that allow passage of the toxic drug metabolites.

Graft versus Host Therapy of Leukemia

The *HSVtk* construct is also used in another application. Patients with acute leukemia treated by bone marrow transplantation frequently suffer a graft versus host reaction as lymphocytes in the donor marrow attack the tissues of the host. While this is a significant source of morbidity in transplant patients, the graft versus host reaction is also very effective at killing any residual leukemic cells. Using this observation, an experimental therapy has been created in which donor lymphocytes are pretreated with *HSVtk* and then infused into leukemic patients. A graft versus host reaction occurs. After a period of days, the reaction is stopped by giving ganciclovir to kill all of the infused donor lymphocytes. This stops the undesirable effects of the reaction after the killing of the patient's leukemic cells has occurred.

Molecular Production of Drugs

Drugs have been harvested from natural sources, usually plants as a tea (water soluble) or as an elixir (alcohol soluble) extract. As new drug discovery turns to synthetic molecules, we need new methods of production of drugs. Let's consider a few classes of drugs and how they can be produced.

Antisense

Antisense drugs are short pieces of DNA that are easily synthesized chemically in an automated process. We have not yet found the optimal way to modify the DNA backbone of antisense molecules to make them more suitable. We need to stabilize the backbone so that the drug is not enzymatically degraded in the blood. Ideally, we would like an oral form that resists digestion and is absorbed. If these problems cannot be solved, antisense may have to be delivered by a vector, becoming a DNA vaccine rather that a drug. Whatever the final answers to the delivery problem for antisense, the great advantage remains: antisense is very specific against a unique target. Synthesis of different antisense molecules is all done by the same process. Changing production from one antisense drug to another is simply throwing a few switches on an automated machine. Throughout this book we have seen potential applications of antisense therapy in many areas, particularly infectious disease and cancer. Once the details are worked out, a huge class of molecules, each with specific targets will be available as drugs.

Recombinant Drugs

Insulin was the first widely used drug produced by *recombinant* DNA technology. In Chapters 1 and 2 we saw how relatively simple the *INS* gene is. The gene was inserted in bacterial and yeast expression vectors. Bioreactors are industrial scale culture tubes for growing large amounts of microbes and harvesting commercial quantities of drug. Before insulin was produced as a recombinant molecule, it was extracted from animal pancreases collected in slaughter houses. Given the current uncertainty about transmission of prion elements from animals to human (recall Chapter 4), imagine the concern we would have if animal-derived insulin were still in widespread use.

Human growth hormone is another protein now available as a recombinant drug. Early human growth hormone therapy used protein extracted from cadaver pituitary glands. Cadaver derived drug was in very limited supply. Good thing, since we now know that Creutzfeld-Jakob prion disease was transmitted to some who received this product. Recombinant drugs remove concern about transmissible infectious agents that were a constant threat from human- or animal-derived products. Antihemophilic factors VII and VIII transmitted HIV and hepatitis B and C to large numbers of patients. These proteins are now available in recombinant form, free of human virus contaminants. Table 8.1 lists a few more examples of recombinant produced drugs.

Pharmed Drugs

A *pharmed* drug is just a recombinant drug taken one step further (see **pharming** in Glossary). Bacteria and yeast will not synthesize more complex proteins. Plants and animals will. **Genetically modified (GM)** plants and animals used to produce drugs are in essence just very large expression vectors. Of course, a pharm looks very different from a bioreactor vessel, but the principles are the same. In Chapter 4, we discussed the hepatitis banana and the rotavirus potato. These GM plants express antigens that, when ingested, invoke some level of immunity against the foreign protein cloned into their genome. They are potential oral vaccines, and examples of pharmed drugs. Let's look at a few other early clinical uses of GM plants.

Table 8.1. Examples of recombinant and pharmed drugs.

Recombinant	Pharmed
EPO	Soy—antihypertensive
GM-CSF	Milk—TPA
Factors VII and VIII	Tobacco—TPA
Insulin	Banana—hepatitis vaccine
Interferon	Potato—rotavirus vaccine
Interleukins	
TPA	

EPO, erythropoietin; GM-CSF, granulacyte/macrophage colony-stimulating factor; TPA, *tissue plasminogen activator.*

Miniexamples

CEA Antigen in Cereals

An antibody against carcinoembryonic antigen (CEA) has been produced in a scaled-up production model using rice and wheat as the host plants. A single-chain Fv antibody (intrabody) against CEA could be harvested from dried plant material at a concentration of tens of micrograms per gram of plant. The protein was stable in dried plant material for long periods. This pharming of a possible anticancer antibody as a stable protein in a production crop goes beyond earlier research applications of growing a human protein in a laboratory-raised transgenic plant.

Spider Silk

The silk produced by spiders has long been known to be incredibly strong, light and elastic. Unfortunately, spiders don't farm very well. If we can't herd arachnids, at least we can now pharm their genes. Expression of silk protein genes (also called spidroins) has not worked in bacteria. Bacteria, it turns out, do not like to make the protein; they do not have enough of certain amino acids. Tobacco plants are a better expression system. Transgenic tobacco plants with spider silk genes have been engineered. Up to 2% of the total protein is spider silk, a yield ready for commercial processing. Spider silk in commercial quantities has many potential uses, such as a recyclable substitute for plastics. Biomedical applications need a very lightweight strong material of organic origin; this may be one answer.

Aside: Franken Foods

The controversy over the use of transgenic plants in agriculture, "franken foods," as they have been called by detractors, will likely be fought for some time to come. The battle will involve politics, economics, marketing (both pro and con), and a general public sense of what the genome entails. The technical advantages of transgenic plants are many: decreased use of pesticides, more efficient use of water, higher and quicker yields, and crops that grow on poor land. Going beyond these advantages, transgenic crops will soon offer nutrition not previously available. A gene that adds iron to rice makes it the almost-complete food. The possibility even exists of adding animal proteins to a plant genome, sidestepping the food chain to achieve "beef" without a cow eating grass.

The negative side of transgenic foods is the unknown. A single very successful strain of rice adds to the problem of monoculture. All of our food eggs would be in one basket if a disease struck that one strain of rice. Some worry about the equivalent of a Y2K time bomb inadvertently programmed into a plant genome. New plants have unknown ecological impact. In fact, any change in agriculture alters our environment. New plant proteins may evoke new and unexpected allergies. A transgenic plant is a commercial product; the company that produces it is liable for bad effects. Preexisting, natural food products are exempt. A patient who cannot eat wheat because of gluten-sensitive enteropy has no one to sue. A patient whose allergy to eggs is worsened by eating a new transgenic cereal may have a suit.

I have spent some time trying to come to a personal opinion about the correctness of transgenic foods, and I am frequently asked what I think. At the moment, I have not decided. I do know that my concern and that of the Western world may be moot. I am well fed and can afford to say no to transgenic corn. However, a poor country that finds that it can improve its food supply may not stop to ask the questions that have been raised. Hunger has always proven itself to be a powerful force.

Biomaterials and Tissues

The transplantation of tissues and organs is severely limited by the availability of donors, and secondarily by the problems of immune rejection. The engineering of tissues made from stem cells grown on an artificial matrix may provide an alternative. Skin, cartilage, and bone have been the most studied. The extracellular matrix of all three of these organs induces stems cells to differentiate into the appropriate tissues. Eventually the cells produce extracellular matrix of their own and the original supporting artificial scaffold is resorbed. As we learn more about different types of stem cells and gather experience with culturing them, the engineering of tissues may expand to include entire organs. Cytokines, now widely available as recombinant proteins, further serve to induce tissue and organ growth and differentiation, increasing the types of tissues that can be produced.

Table 8.2. Sources of stem cells.

Tissue	Potential
Blood, bone marrow	Pluripotent hematopoietic stem cells can regenerate blood and immune system
Placental cord blood	Naive pluripotent hematopoietic stem cells; can be transplanted into non-HLA matched recipients
Placental tissue and fat	Possible totipotent stem cells; range of tissues types uncertain
Any adult cell	In theory, with reprogramming and dedifferentiation, can be made into totipotent stem cell, as in cloning
Embryonic stem cells (obtained from fertilized zygotes in aborted tissue, or from discarded IVF embryos; possibly can be raised in cell culture)	Totipotent stem cells, all tissue types

HLA, human leukocyte antigen; IFV, in vitro fertilization.

Embryonic Stem Cells

In Chapter 3, we considered embryonic stem (ES) cells as a possible starting point for cloning. Here, we need to look at ES cells as a very useful resource for tissue replacement or regeneration. ES cells may be derived from a blastocyst at about 5 days after formation of the fertilized zygote. At this stage, lack of differentiation gives ES cells the potential to differentiate into almost any tissue in the body. They can also be proliferated in cell culture, generating a large number of cells for use from an initial few. Human ES cells can be recovered from embryos created for in vitro fertilization, after the use of these embryos by the parental couple is completed. Many see this as an ethical source of ES cells (see Guenin, 2001). Another source of ES cells is from early abortions, a source felt to be unethical and probably illegal. ES cells can be injected into tissues and stimulated to grow and differentiate, regenerating function of organs. There is a potential role for tissue regeneration in heart, liver, brain, spinal cord, cartilage, and muscle. The controversy surrounding the use of ES cells, even for research, will likely continue for some time. Table 8.2 summarizes the sources of stem cells for all types of tissue transplantation and regeneration. Blood- or marrow-derived pluripotent stem cells and placental cord blood stem cells are used for transplantation of the blood and immune systems. These cells, and the other sources listed in Table 8.2, have limited ability to regenerate tissues, but may serve if ES cells are found to be an unethical or illegal source. The issue of ES cells is the focal point of the current ethical dilemma posed by genetic engineering and cloning in general.

Summary

Molecular therapy is derived from our knowledge of the human genome and our ability to manipulate it. Pharmacogenetics is a new way of looking at how a patient and a drug will likely interact. We have the potential of using the idiosyncratic response of a person to a drug as an advantage rather than being surprised by it as a side effect. Computer modeling and screening of genomic-drug data is revolutionizing the way we search for new drugs. Drugs were previously derived from plant extracts discovered by trial and error. New drugs will be synthetic molecules produced to interact with a specific target. The methods of drug production also have changed. We now use genes inserted in bacteria or yeast to make recombinant drugs, or going further, genetically modified plants and animals to make pharmed drugs. Gene therapy goes beyond what we are used to considering as a drug. The goal of gene therapy is to insert the correct information into the body. Ideally, this is far superior to drugs; we are correcting the fault rather than attempting to override it.

Bibliography

Goldman JM, Melo JV. Editorial. Targeting the BCR-ABL tyrosine kinase in chronic myeloid leukemia. *N Engl J Med* 2001;344:1084–1086.

Guenin LM. Morals and primordials. *Science* 2001;292:1659–1660.

Mannucci PM, Tuddenham EGD. Review article. The hemophilias—from royal genes to gene therapy. *N Engl J Med.* 2001;344:1773–1779.

McCarthy JJ, Hilfiker R. The use of single-nucleotide polymorphism maps in pharmacogenetics. *Nature Biotechnol* 2000;18:505–508.

Murray M. Mechanisms and significance of inhibitory drug interactions involving cytochrome P450 enzymes. *Int J Mol Med* 1999;3:227–238.

Roses AD. Pharmacogenetics and the practice of medicine. *Nature* 2000;405:857–865.

Scheller J, Gunrs K-H, Grosse F, Conrad U. Production of spider silk proteins in tobacco and potato. *Nature Biotechnol* 2001;19:573–577.

Shi MM. Enabling large-scale pharmacogenetic studies by high-throughput mutation detection and genotyping technologies. *Clin Chem* 2001;47:164–172.

Wall RJ, Kerr DE, Bondioli KR. Transgenic dairy cattle: genetic engineering on a large scale. *J Dairy Sci* 1997;80(9):2213–2224.

9   Conclusions

We have covered a lot of subjects in a short book. We began by looking at DNA as a million miles of railroad track where every cross-tie needed to be examined. The Human Genome Project completed that task for us. Next we envisioned the DNA in the cell nucleus as a basketball containing kilometers of spiderweb. (Later we found out that spiderweb protein has been cloned, so we can make our basketball model if we need to). The information content of the genome was then compared to a library. Initially this library was "read-only." But we soon found out that it is easy to manipulate DNA. We may, if we choose, alter our own genomes and write into the library that contains our genetic information.

Our first clinical examples of the use of molecular medicine were in infectious diseases. Modern therapy of infections involves novel ways to attack microbes with DNA vaccines, ribozymes, intrabodies, and all sorts of fabricated molecules. Understanding the genome of microbes makes us realize how easily they can mutate to adapt to a changing environment. Bioterrorism is an unfortunate side effect of appreciating the molecular basis of virulence in infectious diseases.

We have explored the basic genetic mechanism of mutation and DNA repair. Simple genetic diseases like factor V Leiden are due to a single base pair switch that is easy to detect; we may someday screen everyone for this. Polygenic diseases, including diabetes mellitus and atherosclerosis, are much more complex. The genetic component of these diseases is important, but not everything. The environment also shapes how the disease will develop. The human genome does not code for everything; in many instances it only sets the stage. The immune system was another example of this. The immune system depends on the shuffling of the immunoglobulin and T-cell receptor genes to allow for great diversity in antibody production. But the immune system also develops based on its exposure to antigen, and on what it remembers and what it learns from those exposures.

Cancer is, at the core, entirely a molecular disease. Cancer results from multiple mutations in the cell signaling pathway (oncogenes) and additional mutations in the DNA repair system (tumor suppressor genes). This knowledge is the basis of new early detection methods and new therapies including cancer vaccines. We hope we will be able to shut down abnormal oncogene function with specific antisense therapy.

Molecular therapies are the newest product and most dramatic result of understanding the molecular basis of disease. Gene transplantation to fix specific defects in inherited diseases has begun. We expect to be able to do much more in the near future. We even have the possibility of improving on the human genome. First steps will be in utero trans-fection with DNA vaccines to prevent infectious diseases. Later steps might include limiting some of the programmed steps in aging.

Finally we find ourselves at the frontier of a revolution called molec-ular medicine. We have the tools to clone ourselves, to alter our genetic heritage, and to create new plants and animals to serve us. The tools are, for the moment, far more advanced than our knowledge and expe-rience. This book has explored our knowledge of the molecular basis of life, to prepare us for its effect on medicine.

Glossary

allele: one of possible multiple forms of a gene that can occupy the specific chromosomal location for that gene.

angiogenesis: the process of inducing blood supply to new tissues; a factor in tumor growth.

antisense: a short segment of DNA, synthesized to be complementary in genetic sequence to the sense strand of RNA; used to block gene translation.

apoptosis: (Greek, "fallen leaves") the process whereby a cell commits programmed cell death in response to internal signals.

carcinoma in situ: a stage in the multistep evolution of cancer in which malignant cells are present but without invasion.

chimera: an organism with additional genetic material beyond its own germline genome

clone: a population of genetically identical cells.

cloning: the process of copying a gene or an organism.

codon: three base pairs of DNA code for a single amino acid during the translation of a gene into protein.

crossing over: reciprocal breaks in homologous chromosomes at meiosis resulting in the re-sorting of genetic material.

DNA vaccine: a new class of vaccines based on DNA inserting into the cell; the intracellular translation of this DNA produces major histocompatibility complex (MHC) type I immunity; a possible vaccination method for human papilloma virus (HPV) and human immunodeficiency virus (HIV).

dominant: a gene that when present as only one copy out of two alleles produces sufficient protein to have an effect; the opposite of recessive.

embryonic stem cell: a cell taken from the early blastocyst of a fertilized zygote; capable of regeneration into an organism and differentiation into many different tissues.

exon: a segment of DNA within a gene that codes for protein, separated from adjacent exons by noncoding segments called introns.

expression array: a DNA on a chip microarray that detects the messenger RNA (mRNA) molecules currently being produced by a group of cells, giving an indication of cell function.

fluorescent in-situ hybridization (FISH): a method for analysis of chromosomes and genes, using fluorescent DNA probes.

gene: a piece of DNA the encodes a packet of information that is part of a cell's permanent structure and is copied into daughter cells at division.

genetically modified (GM): an organism that has had its germline DNA modified; usually carried out in a plant to produce an improved strain.

genomics: the study of how genes function.

germ cell: the oocyte or sperm; also sometimes refers to fertilized zygote.

germline: the genome of an organism at birth.

in vitro fertilization (IVF): the combining of sperm and oocyte to form a zygote outside the body, following by implantation.

intrabody: a synthetic single-chain antibody produced by genetic engineering.

intron: a segment of DNA within a gene that does not codes for a protein, occurring as a space between exons.

knockout mouse: a genetically modified mouse, engineering to lack a specific gene function.

microarray: a distribution of gene probes on a substrate, used for rapid large scale genetic analysis.

oncogene: a growth control gene, part of the cell signaling pathway. The mutated form of an oncogene is a step in formation of a tumor.

open reading frame (ORF): a long segment of DNA that when read as triplets, produces a set of codons that does not include a STOP codon. An ORF is a sign that this segment of DNA is part of a gene.

pharmacogenetics: the study of an individual's response to drugs based on that person's genome.

pharming: the process of creating new molecules by growing them in genetically altered plants or animals.

polygenic: a condition that depends on multiple genes.

polymerase chain reaction (PCR): a method for amplifying pieces of DNA; used as a detection method and to produce large amounts of specific DNA fragments.

posttranslational modification: a modification in protein structure that occurs after the protein is formed by translation on the ribosomes.

prion: an abnormal protein, infective in that it is capable of causing the transformation of other proteins.

promoter: a molecule that stimulates gene function by binding to DNA.

proteomics: the study of proteins carried out by looking at genome information; the structure and possible function of a protein is inferred from genetic sequences.

recessive: when only one copy of a recessive gene is present, no abnormality is expressed, the opposite of dominant.

recombinant protein: a protein produced by introducing a gene into a bacteria or yeast (see pharming).

restriction enzyme: an enzyme, derived from bacteria that cuts DNA at a specific site based on nucleotide sequence; a major tool of recombinant DNA technology.

ribozyme: an RNA molecule that has enzymatic activity; usually in cleaving other RNA molecules.

single nucleotide polymorphism (SNP): single base pair differences in the DNA, about 2.5 million occurring in the human genome; these alterations characterize the difference in the genetic makeup of one person compared to another.

somatic cell: all cells in the body, besides germ cells.

telomere: a structure at the end of chromosomes that shortens with each progressive cell division.

transcription factor: a molecule that binds to DNA, affecting the function of a gene; see also *promoter.*

transgenic: an organism whose DNA has been altered by biotechnology.

tumor suppressor gene: a DNA repair gene that when mutated allows tumors to develop.

vector: a means of carrying a cloned piece of DNA, usually capable of inserting into a bacteria or cell; typically a bacteriophage or virus.

Index